German G-Type Bombers of WWI

A Centennial Perspective on Great War Airplanes

Jack Herris

Great War Aviation Centennial Series #14

This book is dedicated to my friend and colleague James Davilla, talented MD and aviation historian, for his support, generosity, and contributions to aviation history.

Acknowledgements

My sincere thanks to Greg VanWyngarden, Colin Owers, Bill Toohey, Lance Bronnenkant, Tobias Weber, Dave Hooper, Dave Watts, and Richard Andrews for photographs, Bob Pearson for his color profiles, Aaron Weaver for photographs and image editing, Marty Digmayer for scale drawings, and the Museum of Flight in Seattle for photographs. Thanks also to Jim Dietz for the cover painting and Aaron Weaver for the cover design. I also want to thank Reinhard Zankl for his helpful information regarding production orders and serial numbers and for photographs. Any errors are my responsibility.

Color aircraft profiles © Bob Pearson. Purchase his CD of WWI aircraft profiles for $50 US/Canadian, 40 €, or £30, airmail postage included, via Paypal to Bob at: **bpearson@kaien.net**

Cover painting *Night Raiders* by James Dietz. Please see Jim's website at: **www.jamesdietz.com**

For our aviation books in print and electronic format, please see our website at: **www.aeronautbooks.com**. I am looking for photographs of the less well-known German aircraft of WWI. For questions or to help with photographs you may contact me at **jherris@verizon.net**.

Interested in WWI aviation? Join The League of WWI Aviation Historians (**www.overthefront.com**) and Cross & Cockade International (**www.crossandcockade.com**).

ISBN: 978-1-935881-26-1
© 2014 Aeronaut Books, all rights reserved
Text © 2014 Jack Herris
Design and layout: Jack Herris
Cover design: Aaron Weaver
Digital photo editing: Aaron Weaver & Jack Herris

www.aeronautbooks.com

Table of Contents

German G-Type Bombers	3
AEG Bombers	6
AEG K.I	8
AEG G.I	9
AEG G.II	10
AEG G.III	20
AEG G.IV	26
AEG G.V	36
Albatros Bombers	42
G.I	42
G.II	43
G.III	45
Daimler Bombers	55
Union G.I (R.I)	55
Daimler R.II	59
Daimler R.I	60
Daimler G.I	62
Daimler G.II (480-1915 Tractor)	62
Daimler G.II (480-1915 Pusher)	65
Daimler G.III (584-1916)	67
Friedrichshafen Bombers	70
Friedrichshafen G.I	70
Friedrichshafen G.II	74
Friedrichshafen G.III / G.IIIa	79
Friedrichshafen G.IV	90
Friedrichshafen G.V	94
Gotha Bombers	96
Gotha G.I	97
Gotha G.II	106
Gotha G.III	110
Gotha G.IV	116
Gotha G.V / G.Va / G.Vb	125
Gotha G.VI	142
Gotha GL-Types	144
Gotha G.VII / GL.VII	144
Gotha G.VIII / GL.VIII	150
Gotha G.IX / GL.IX	152
Gotha G.X / GL.X	157
Halberstadt G.I	160
Hansa-Brandenburg G.I	161
LVG Bombers	165
G.I	165
G.III	165
Roland G.I	168
Rumpler Bombers	170
Rumpler A415	170
Rumpler G.I	172
Rumpler G.II	178
Rumpler G.III	184
Schütte-Lanz G.I	187
SSW L.I	190
G-Types in Retrospect	192
Bibliography	192
Index	192
Scale Drawings (1/72 – MD)	
Albatros G.III	193
Rumpler G.I & G.II	196
Rumpler G.III	199
Afterword	202

Above: AEG G.II G.7/15 of *FA* 42 with pilot *Offz.* Edmund Nathanael in the center of the aircrew. Nathanael received the Wilhelm Ernst War Cross and other awards with *FA* 42. After transfer to *Jasta* 5 he scored 14 victories before being KIA.

Preface

The purpose of this book is to document German multi-engine bombers of WWI other than the true R-types (*Riesenflugzeug*), or giant bombers. This book primarily covers the German G-type bombers, mostly twin-engine aircraft, and also includes several other types in a similar class. For example, the SSW L.I is included because, despite being a larger, three-engine type, it was intended for a similar purpose and was the only L-type completed before the Armistice. Similarly, the Daimler R.I and R.II, developments of the Union G.I (also known as the Union R.I), are included despite their designation as R-types because they were not true R-types. That is, their engines were not serviceable in flight, which was a requirement for a true R-type, and although they had four engines they were generally in the same class as the typical twin-engine G-type due to their low engine power. Similarly, the Union G.I was not a true R-type despite its alternate R.I designation.

German G-Type Bombers

Early in the war Germany pursued the battleplane or 'Kampfflugzeug' concept, the specifications for which were determined before the war, without the benefit of combat experience, and promulgated in July 1914. These specifications led to development of a number of twin-engine aircraft in Germany, the first of which to be completed was the AEG K.I, the 'K' standing for 'Kampfflugzeug'. The *Kampfflugzeug* was envisioned as a sort of aerial cruiser with gunners engaging enemy aircraft with flexible guns in a manner analogous to naval combat.

War-time operational experience quickly revealed that the slow *Kampfflugzeug* was better suited to bombing than intercepting other aircraft and the designation of the AEG K.I was changed to AEG G.I, 'G' standing for '*Grossflugzeug*', literally 'large aircraft' but in this case also having the connotation of twin-engine bomber.

When the *Luftstreitkräfte* (Air Service) was reorganized in October 1916, four *Kagohls* (*Kampfgeschwaders*) were disbanded and the three remaining were equipped with G-type bombers. *Kagohl* 3 was soon reestablished for attacks on the UK, and nominal bomber strength was 144 G-types organized in four *Kagohls* of six *Kastas* each. In Oct. 1917 the *Kagohls* were redesignated *Bogohls* (*Bombengeschwader der Obersten Heeresleitung* – Bombing Squadron of the Army High Command) each with three *Bombenstaffeln* (*Bostas*) except *Bogohl* 3 that had six for UK raids. From late 1917 G-types performed night bombing only.

The AEG G-types were successful tactical bombers from their appearance at the front. Friedrichshafen G-types were also successful, both they and the AEG series being superior to the Gotha in all respects except range and ceiling. However, neither series became famous because they were used only for tactical bombing behind the lines.

Frontbestand Inventory of G-Type Aircraft (Twin-Engine Bombers) at the Front

Manufacturer and Type		1914 31 Aug	1914 31 Oct	1914 31 Dec	1915 28 Feb	1915 30 Apr	1915 30 Jun	1915 31 Aug	1915 31 Oct	1915 31 Dec	1916 28 Feb	1916 30 Apr	1916 30 Jun	1916 31 Aug	1916 31 Oct	1916 31 Dec	1917 28 Feb	1917 30 Apr	1917 30 Jun	1917 31 Aug	1917 31 Oct	1917 31 Dec	1918 28 Feb	1918 30 Apr	1918 30 Jun	1918 31 Aug
AEG	G.I							1	5		5															
	G.II						2	5	10	13	12	2	4	4	4	2	2	1	1							
	G.III												6	16	22	21	22	9								
	G.IV									1								5 3	9	15	15	35	37	54	74	51
	G.																									
Albatros	G.II																1	9	1	1		1				
	G.III																		2	1	2					
Friedrichshafen	G.II															1		4	8	17	17	10	9	2	1	1	1
	G.III																		9	32	24	57	69	96	74	24
	G.IIIa																								18	95
	G.IV																								4	8
	G.IVa																								5	6
Gotha	G.I							5	6	1		1	1	1												
	G.II													4	3	1	1									
	G.III													7	14	3	4	3	3							
	G.IV														1		30	36	34	35	19	10	8	6	5	
	G.V																	3	20	33	34	36	15	8		
	G.Va																				11	19	4			
	G.Vb																						21			
Rumpler	G.I									1		1														
	G.II									1	7	8	4	1	1											
	G.III												1		3	5	5	10	1	4						
Total:					2	6	20	20	13	7	13	28	48	46	34	71	86	111	116	155	156	206	216	223		

Above: Of all the operational types, those designed by Albatros were built in the smallest quantities and served at the front for the shortest period. Contrary to the *Frontbestand* shown above, the Albatros G.II remained a single prototype. Only the Albatros G.III was produced and served at the front, and the inventory of G.II aircraft shown above were actually Albatros G.III aircraft; the numbers in the two rows should be combined under G.III. The AEG G.IV listed in December 1915 was a transcription error; the AEG G.IV did not reach the front until 1917.

Both the Rumpler and Albatros designs were mediocre and were built in small numbers. Several other manufacturers were even less successful, Daimler in particular being a disappointment, especially considering the valuable engineering resources wasted on the completely unsatisfactory bombers designed by Karl Schopper.

In early 1918 Gotha abandoned their existing designs in favor of smaller, faster designs, but despite prolonged development these did not reach the front. At this time new manufacturers entered the field with designs to new configurations, the LVG G.III and SSW L.I, but despite their promise neither was in time for production before the Armistice.

The German strategic bombing campaign against Britain – first by Zeppelins, then by Gothas, and finally including Zeppelin-Staaken Giant bombers – was a thorn in the Allied side but was never more than a distraction with no hope of decisive success. The technology was simply unable to deliver enough bombs with enough accurately, which precluded any conclusive victory.

However, German G-types were very effective tactical night bombers and the Armistice specifically required the Germans to surrender all G-types to the Allies, who were astonished at the small number of aircraft they received given the damage they had done. As with German aviation in general, resource shortages, especially insufficient engine production, limited the number of G-types Germany could build.

Above: AEG built a wide variety of warplanes but is probably best known for its twin-engine bombers. Above is G.V G.625/18, an example of the final production AEG bomber. The G.V was based on the G.IV with its wing enlarged in span and area for more lift to carry the desired 1,000 kg bomb load, requiring an additional bay of struts. A 'box' tail gave better control during asymmetric thrust after failure of an engine, and the Flettner tabs on the ailerons reduced control forces, making the large bomber easier and less tiring for the pilot to fly. Flettner tabs were also used on late-war Gotha and Friedrichshafen designs. Like most late-war G-types, the G.V was powered by two 260 hp Mercedes D.IVa engines.

Above: Friedrichshafen G.II(Daim) 625/16 was the first Friedrichshafen bomber built under license by Daimler, which produced 298 Friedrichshafen bombers under license. Here it is undergoing its type test at Adlershof in April 1917.

Above: This appears to be the prototype Albatros G.III after the ailerons were fitted with aerodynamic balances to reduce the control forces. The G.III was streamlined for a 1916 two-engine bomber. The thick airfoil section for high lift, unusual for an Albatros design, is clearly evident.

AEG Bombers

Despite the wide variety of AEG aircraft that served operationally, AEG is perhaps best known for its twin-engine bombers. These originated from development of a 'battle plane' that was essentially an aerial cruiser armed with flexible machine guns and bombs. This aircraft lead to a series of designs that evolved into twin-engine bombers based on operational experience.

AEG bombers differed from their competitors in two significant ways. First, their airframes were constructed of welded steel tubing. Second, they used tractor propellers instead of pusher propellers.

Facing Page: An aviator poses with flamboyantly-marked AEG G.III G233/15. The large, four-blade propellers are to harness the power of the 220 hp Mercedes D.IV straight-eight cylinder engines. These rare engines were a transitional types created by adding two more cylinders to the widely-used Mercedes D.III six-cylinder engine. When the powerful, 260 hp Mercedes D.IVa six-cylinder engine became available, production of the D.IV was stopped. Nominally more powerful, a key advantage of the D.IVa engine was its greater reliability, the long crankshaft of the D.IV engine sometimes fracturing in flight.

AEG G-Type Production Orders		
Serial Numbers	**Qty**	**Order Date & Notes**
AEG G.II (27–28 Total)		
G.2–7/15	6	April 1, 1915
G.19–30/15	12	May 6, 1915
G.46–51/15	6	September 7, 1915
Unknown	(3–4)	Sept. 22, 1915 (note 1)
AEG G.III (46 Total)		
G.8/15	1	April 1, 1915 (note 2)
G.52–56/15	5	Sep. 7, 1915 (note 3)
G.210–239/15	30	Dec. 1, 1915 (note 4)
G.143–152/16	10	March 1916 (note 4)
AEG G.IV (320 Total)		
G.153–192/16	40	March 1916 (note 5)
G.1095–1144/16	50	December 1916
G.560–609/17	50	September 1917
G.844–893/17	50	December 1917
G.545–619/18	75	April 1918
G.1215–1264/18	50	July 1918
AEG G.IVk (5 Total)		
G.500–505/18	5	March 1918
AEG G.V (50 Total)		
G.620–644/18	25	April 1918
G.1700–1724/18	25	October 1918
Notes: 1. 12 ordered but only 3–4 delivered 2. Prototype, single rudder, serial unconfirmed. 3. G.53/15 had triple rudders 4. Single rudder 5. Includes some AEG G.IVb 6. The AEG G.I serial was G.1/15.		

Above: An aviator poses with his AEG G.II. The AEG G-types all followed the same basic configuration; conventional biplanes with steel-tube structures covered by fabric and two engines mounted as tractors. Nearly all other German bombers had engines mounted as pushers.

Below: AEG G.IV 157/16 of *KG* 4 after an exciting landing at *AFP*4 at Ghent on August 11, 1917. Any landing you can walk away from…

AEG G-Type Specifications					
	G.I	**G.II**	**G.III**	**G.IV**	**G.V**
Engine	2x100 hp Mercedes D.I	2x150 hp Benz Bz.III	2x220 hp Mercedes D.IV	2x260 hp Mercedes D.IVa	2x260 hp Mercedes D.IVa
Span Upper	16.00 m	16.20 m	18.44 m	18.40 m	27.24 m
Span Lower	15.20 m	15.20 m	17.20 m	17.40 m	26.30 m
Chord Upper	2.20 m	2.20 m	2.50 m	2.40 m	2.80 m
Chord Lower	2.20 m	2.20 m	2.50 m	2.40 m	2.39 m
Gap	2.30 m	2.30 m	2.60 m	2.20 m	3.00 m
Wing Area	61.0 m²	61.0 m²	74.0 m²	??? 68.7 m²	127.2 m²
Length	8.7 m	9.1 m	9.20 m	9.70 m	10.80 m
Track	3.15 m	3.15 m	2.85 m	5.10 m	4.85 m
Empty Weight	1,160 kg	1,450 kg	1,940 kg	2,400 kg	2,700 kg
Loaded Weight	1,610 kg	2,050 kg	2,560 kg	3,635 kg	4,800 kg
Maximum Speed	125 kmh	140 kmh	150 kmh	165 kmh	145 kmh
Climb, 1000m	—	11 min.	6 min.	5 min.	6 min.
Climb, 2000m	—	—	—	11 min.	12 min.
Climb, 3000m	—	—	—	21 min.	23 min.
Climb, 4000m	—	—	—	40 min.	34 min.
Armament	2 flexible machine guns, small bombs	2 flexible machine guns, 200 kg bombs	2 flexible machine guns, 240 kg bombs	2 flexible machine guns, 300 kg bombs	2 flexible machine guns, 1,000 kg bombs
Note: The AEG G.IVb wing span was enlarged to 24 m; this enabled a 1,000 kg bomb to be carried. For short missions up to 1,500 kg of bombs could be carried.					

AEG K.I

In March 1914 the German general staff sanctioned the development of the *Kampfflugzeug* (battle plane), and in July 1914 *Idflieg* issued specifications for the type in preparation for a competition planned for spring 1915. The *Kampfflugzeug* concept was basically an 'aerial cruiser' armed with machine guns and bombs. The aircraft was to have 200 hp, carry a crew of three, and have an endurance of six hours.

AEG responded to the requirement with a biplane powered by two 100 hp Mercedes D.I engines; the factory designation was GZ1 and the military designation was AEG K.I. The K-type designation was soon changed to G-type, the 'G' standing for '*Grossflugzeug*' or large aircraft, later to become synonymous with twin-engine bombing aircraft. The K.I had side-by-side seating for two crewmen and a single flexible machine gun was mounted in a nose turret because the aircraft was intended to chase and destroy enemy airplanes. The airframe was constructed of self-fused (autogenous) welded steel tubes. The nose was covered with light armored plate and the rest of the aircraft was fabric covered.

In January 1915 the AEG K.I prototype was first flown by test pilot Willy Kanitz and gave promising results during flight tests in January–February. This convinced *Idflieg* to order a second prototype as the AEG G.I for combat evaluation.

Above: The AEG K.I designed to the flawed *Kampfflugzeug* concept was the first twin-engine AEG design. All subsequent AEG G-types followed the same basic configuration; conventional biplanes with fabric-covered steel-tube structures with two engines mounted as tractors. Almost all other German bombers had engines mounted as pushers. The K.I was a compact design with good handling qualities.

AEG G.I

The second AEG *Kampfflugzeug* prototype, now designated AEG G.I but internally retaining the company designation GZ1, was completed in March 1915 and may have incorporated components from the K.I. The G.I differed from the K.I primarily in its crew and armament; the G.I had three crewmen and flexible machine guns in fore and aft turrets.

The AEG G.I was shipped from the factory on April 24, 1915 for the 4.*Armee* for combat assessment without first being tested at Adlershof. Only one G.I was built.

Above: The AEG G.I differed from the K.I primarily by having an additional crewman with a flexible gun mounted aft. The G.I may have incorporated some components of the K.I.

Right: The nose of the AEG G.I opened to show details of the front gunner's cockpit.

AEG G.II

Prior to the operational trials of the AEG G.I, *Idflieg* ordered six aircraft of an improved type, the AEG G.II (factory designation GZ2). *Idflieg's* requirements for the AEG G.II included two 150 hp Benz Bz.III engines, a crew of two with three seats, a single machine gun, a 200 kg bomb load, and 150 kg of armor plate (front and side 2.3 mm and floor 1.5 mm). Controlling a multi-engine aircraft after the failure of one engine was dependent on both pilot technique and design of the aircraft; accordingly *Idflieg* urged AEG "to make every effort to assure that the aircraft would maintain a straight flight path with only one engine running at full power." With the more powerful engines in the G.II a larger fin and rudder were needed to main control with asymmetric thrust.

As expected the more powerful engines gave the AEG G.II better performance and greater load-carrying capability, and *Idflieg* ordered a second batch of 12 G.II aircraft on May 6, 1915 before the results of the combat trials of the AEG G.I were known. The first two G.II aircraft were completed in May and reached the front in June, with a maximum of 13 at the front in December.

The AEG G.II was not only a new aircraft but was exploring a new combat role as a multi-engine *Kampfflugzeug*, or aerial cruiser. The G.II was used both to escort single-engine reconnaissance and bomber aircraft and to attack enemy aircraft.

Below: AEG G.II G.3/15 was the second production G.II and has the single fin and rudder originally used.

Based on operational experience the first six G.II aircraft (G.2–7/15) were extensively modified for both technical and operational reasons, with the result that no two aircraft were alike. The square, armor-plated nose was replaced by a streamlined, unarmored one. Two gravity tanks and new oil tanks were installed. Because the small fin and rudder failed to provide adequate directional control with one engine out, some machines were retro-fitted with triple rudders, becoming standard beginning with the second G.II production batch (G.19–30/15).

Despite all the modifications to the G.II, including adding a second and sometimes a third machine gun, operational experience soon made it clear that the multi-engine *Kampfflugzeug* concept was a failure. The *Kampfflugzeug* had only modest success as a multi-seat escort and was too slow and cumbersome to catch faster, more maneuverable enemy aircraft – and most enemy single-seaters were faster and more maneuverable.

However, bomb racks were installed in all G.II aircraft and aircrews soon discovered that bombing was the most effective role for the G.II. By mid-1916 *Idflieg* summarized the operational record of the G.II saying it "had fared poorly in air combat, but had been successfully employed as a bomber in squadron strength." The AEG G.II thus discovered the true role of the G-type as a bomber. It remained at the front through June 1917 and set the standard for future AEG bombers. A total of 27 G.II aircraft were delivered before production was shifted to the improved G.III in May 1916.

Above: An AEG G.II in flight. This image has also been identified as a G.IV, but the engines and radiators indicate it is a G.II.

Left: AEG G.II G.7/15 with armored nose.

Below: AEG G.II G.19/15 serving with *Flieger-Abteilung* 22 features triple rudders and an unarmored nose. It appears to have a two-color sprayed camouflage.

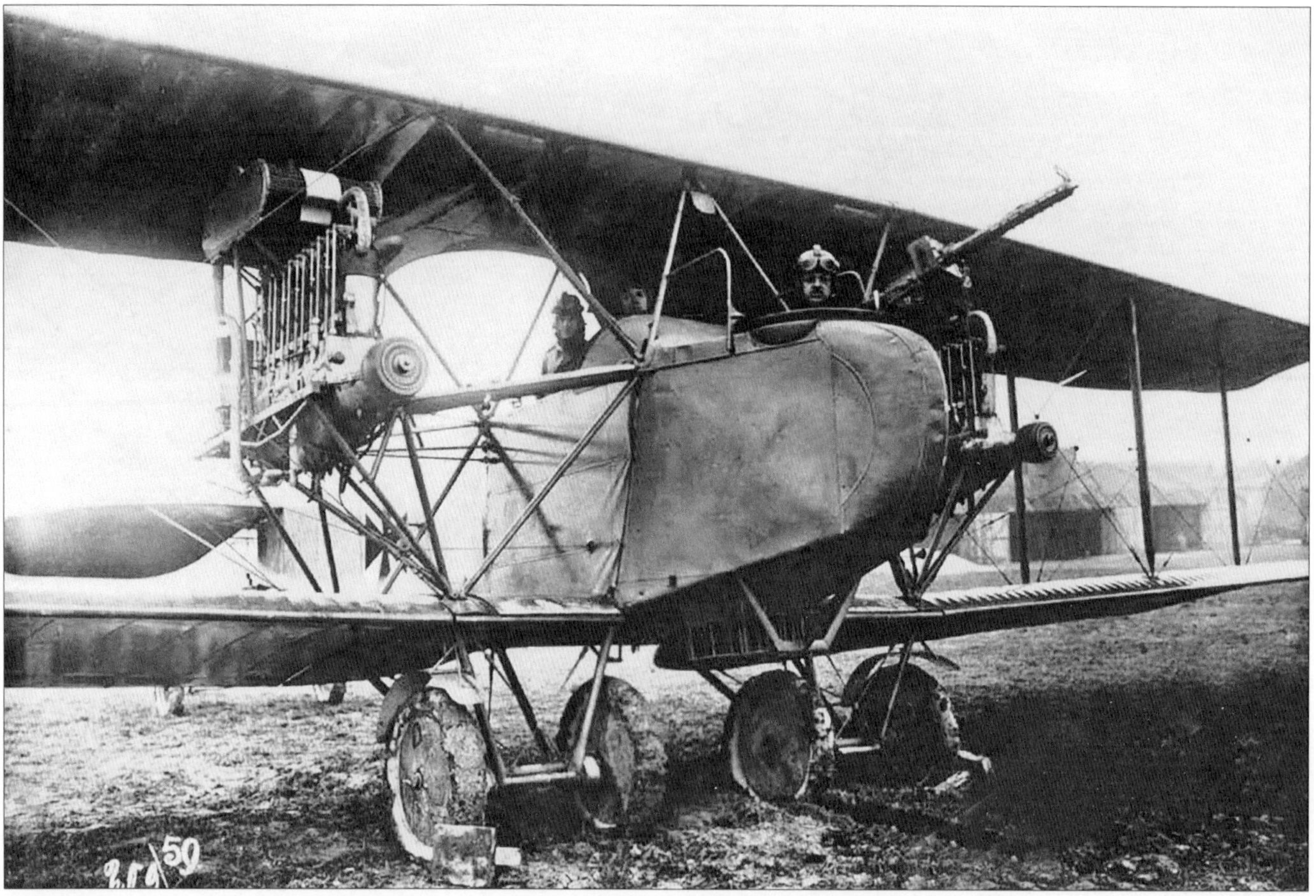

Above: Under-fuselage bomb racks are visible on this AEG G.II without nose armor.
Below: Future ace Rudolph Berthold flew AEG G.II G.21/15 while serving with *FFA* 23.

Above: Rudolph Berthold sits in the cockpit of AEG G.II G.26/15 that he also flew while serving with *FFA* 23.

Below: Rudolph Berthold standing in the cockpit AEG G.II G.21/15 during a royal visit by Duke Ernst August of Brunswick and Prince August Wilhelm of Prussia to *FFA* 23 on October 23, 1915.

Above: AEG G.II with triple rudders and two-color sprayed camouflage finish.

Below: This AEG G.II is at the factory and has outsize triple rudders fitted without vertical fins. This was a prototype; as seen above the production aircraft with triple rudders had a central fixed fin.

Above: AEG G.II with triple rudders and early single color finish.
Below: AEG G.II with triple-rudders and two-color sprayed camouflage finish.

Above & Below: AEG G.II with triple rudders and fairly dark finish. There is a lot of contrast between the overall finish and the white background of the national insignia, especially on the rudders.

AEG G.II

AEG G.II G.4/15.

AEG G.II G.19/15 at *FFA*22, possibly flown by Walter von Bülow.

AEG G.II G.23/15 of *FFA*1 in Salonika carrying the name *Sonnenvogel*.

Above & Below: AEG G.II with single rudder. The single color finish is so light there is almost no contrast between the overall finish and the white background of the national insignia.

Above: AEG G.II with triple rudders and single color finish.

Below: AEG G.II with triple-rudders and single color finish. The color appears darker than some of the earlier monotone finishes used on G.II aircraft like that shown above; on this aircraft there is more contrast between the white background for the national insignia and the overall color. Unfortunately, the serial number is not visible.

AEG G.III

When *Idflieg* ordered the first six G.II on April 1, 1915, AEG was also requested to build a *Kampfflugzeug* with two 220/240 hp engines, the type to be decided later. The intention was to provide greater performance and payload, and the resulting aircraft was designated the AEG G.III (factory designation GZ3). A crew of three, 200 kg of armor protection, and a bomb load of 240 kg were specified. Two machine guns or a cannon mounted in the nose and a rear machine gun was the specified armament.

The G.III was very similar to the G.II although the wingspan was 2.24 m longer. Two of the new, 220 hp Mercedes D.IV straight-eight engines were fitted because these were the most powerful engines then in production. A four-bladed propeller was used to absorb all the power of the engine, but the long crankshaft was subject to fractures in service. The AEG specification chart shows December 1915 as the date the first G.III was completed, but the first three production aircraft were not delivered until May 1916. The G.III prototype was tested with a single fin and rudder and an early production aircraft (G.53/15) was delivered with a triple rudder for comparison. All other production G.III bombers had the single fin and rudder and most had external bomb racks.

The G.III reached the front in June 1916 and was first used as an escort aircraft for single-engine bombers, but it quickly became clear that bombing was the appropriate role for the G.III. *Kampfgeschwader* 1 became "the first formation to be completely equipped with twin-engine aircraft of the G-category for the sole purpose of bombing." Serving with *KG*1 in Macedonia, the G.III was primarily used as a bomber but at least two strafing attacks were made. The last three G.III bombers were delivered in January 1917, bringing total G.III production to 45 aircraft; the AEG G.IV then followed the AEG G.III in production.

Above: AEG G.III G.54/15 serving at the front. The four-blade propellers are a key G.III identification feature.
Below: AEG G.III G.52/15 serving at the front. The G.III was basically an enlarged, more powerful G.II.

AEG G.III

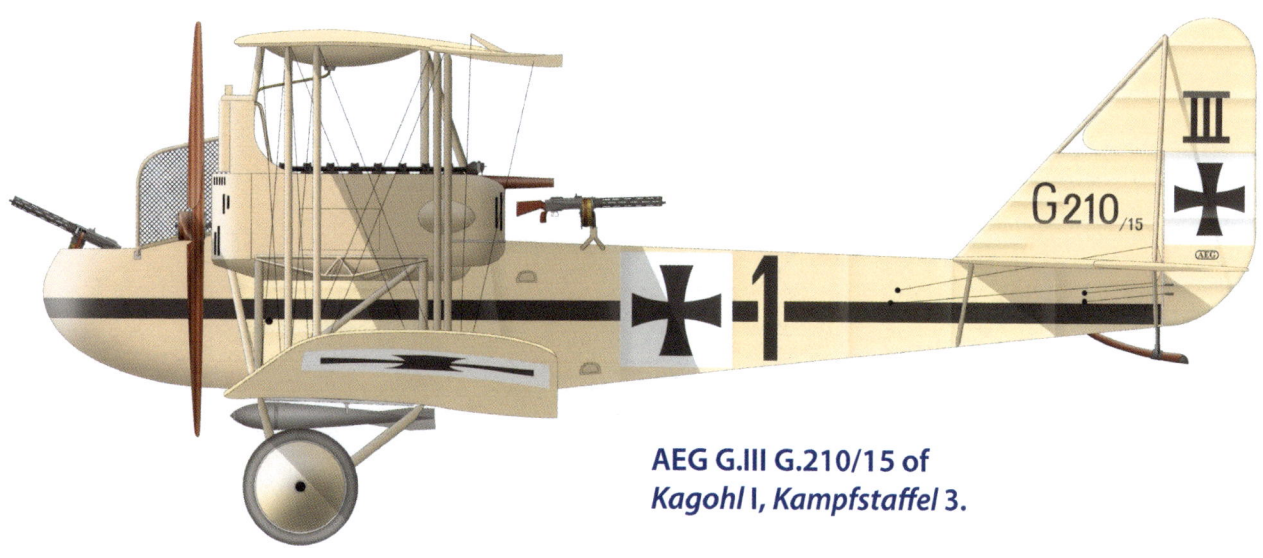

AEG G.III G.210/15 of *Kagohl* I, *Kampfstaffel* 3.

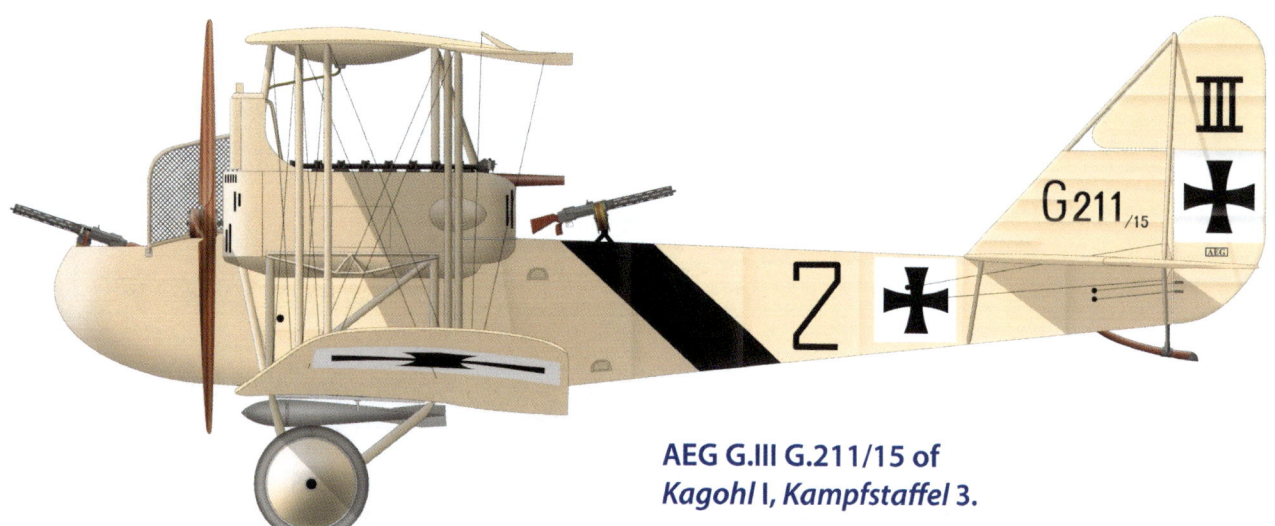

AEG G.III G.211/15 of *Kagohl* I, *Kampfstaffel* 3.

AEG G.III G.213/15 of *Kagohl* I, *Kampfstaffel* 5.

Above: AEG G.III G.233/15; the enlarged rudder with aerodynamic balance helped the pilot maintain control with an engine out despite its more powerful engines. The aircraft wears a very light overall monotone finish.

Below: This portrait of an AEG G.III became Sanke Card 1060. The finish was two camouflage colors sprayed on.

Above: AEG G.III G.213/15 tactical number '3' perhaps serving with *Kasta* 5 based on the Roman numeral on the rudder. The straight-eight cylinder Mercedes had good power but the long crankshaft was subject to failure, especially in multi-engine aircraft.

Below: AEG G.III G.216/15 in overall light finish being repositioned on the airfield.

Above: A mix of light-colored and camouflaged AEG G.III bombers of a *Kampfgeschwader* are lined up.

Above: The aircrew of an AEG G.III bomber flank damage to their aircraft likely caused by anti-aircraft fire. This photo gives a good view of the complex struts supporting the 220 hp Mercedes D.IV engines.

AEG G.III

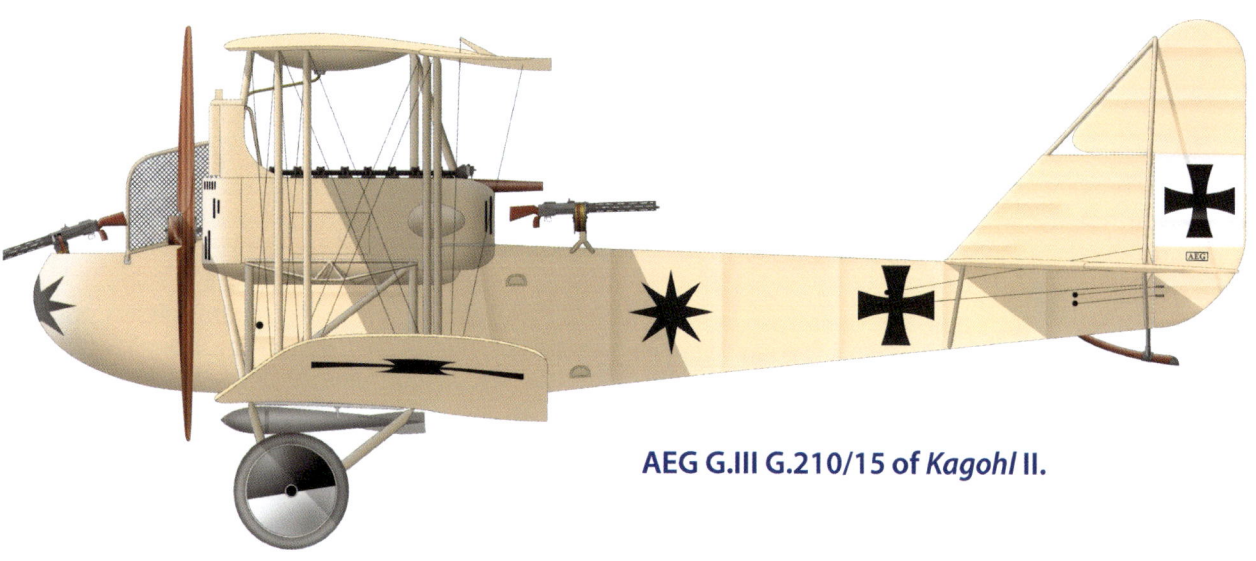

AEG G.III G.210/15 of *Kagohl* II.

AEG G.III G.143/16 seen at the *Geschwaderschule* at Paderborn. Most of the original plain finish has been recovered with night lozenge.

AEG G.III G.152/16 from *Kagohl* I, *Kampfstaffel* 5, during spring 1917 in Salonika.

AEG G.IV

The next step in development of the AEG twin-engine bombers was to give the aircraft more powerful and reliable engines, and the G.IV used the new 260 hp Mercedes D.IVa. The new engine not only had more power but was a six-cylinder engine with more robust crankshaft and was notably more reliable than the earlier D.IV straight-eight.

In March 1916 *Idflieg* placed an initial order for 40 AEG G.IV bombers; the requirements specified a load of 25 x 12 kg or 6 x 50 kg bombs (300 kg total), four machine guns or cannon, a climb to 5,000 meters in 75 minutes, and a top speed of 140 km/h. The G.IV (factory designation GZ4) prototype was completed in September 1916 and exceeded all the requirements except for climb.

The first G.IV production aircraft were delivered in January 1917, immediately after completion of the last G.III aircraft. The G.IV was essentially the same configuration but airframe dimensions differed and the more complex joints were machined from billet steel for greater strength. Its tough steel tube airframe made the AEG G.IV much more robust then the wooden Gotha and Friedrichshafen bombers, especially in crashes. Furthermore, the AEG G.IV was easier to fly than those bombers and did not require the nose-mounted or wing-mounted auxiliary landing gear to prevent nose-overs on landing. For its robustness and better handling the AEG G.IV earned a better reputation amongst German bomber aircrew than its competitors.

Starting in April 1917 the G.IV began to replace the G.III at the front, and served until the end of the war.

AEG G.IV Engine Experiments

Unlike nearly all other German bombers, the engines of AEG bombers were mounted in tractor configuration, not as pushers. AEG experimented with other engines in the G.IV between September 1917 and March 1918, including the 245 hp Maybach Mb.IVa and 300 hp Basse & Selve BuS.IVa engines, to increase the climb rate and payload. Both engines gave better performance than the standard 260 hp Mercedes D.IVa, but the Mercedes remained the G.IV production engine because other aircraft types had priority on the new engines.

In parallel with the experiments with different engine types, experimental turbo-charged Mercedes engines were installed in a G.IV for flight tests. Unfortunately, that aircraft was destroyed in March 1918. Initially Schwade compressors were used, but manufacturing problems at Schwade motivated AEG to build experimental turbo-compressors that were tested in the summer of 1918. In September 1918 *Idflieg* reported that a G.IV bomber equipped with AEG turbo-compressors raised the operational ceiling from 4200 meters to 6000 meters. As a result AEG received a contract to supply 20 turbo-compressors for combat evaluation in bomber aircraft.

Below: Three-bay AEG G.IVb G.168/16 was from the first G.IV production batch and was rebuilt as an extended-span G.IVb.

AEG G.IVa

The AEG G.IVa has not been identified in *Fliegertruppe* records, and it is unknown if it was an un-built project or an actual variant of the G.IV.

AEG G.IVb

In mid-1917 some AEG G.IV bombers from the first production batch were given wings of increased span and area to improve their climb rate and ceiling. These aircraft, fitted with three-bay wings of 24.0 m span, were designated the G.IVb. In September 1917 *Idflieg* decreed that "since the AEG G-types are (now) employed solely for night bombing it is possible to forego the higher climb rate of the three-bay machine in favor of the superior speed and maneuverability of the twin-bay machine. Henceforth only twin-bay aircraft will be built." Given the extra load carrying capability, some G.IVb bombers were modified to carry the 1,000 kg (2,200 lb) P.u.W. bomb.

AEG G.IVb-Lang

In March 1918 an AEG G.IVb powered by two 300 hp Basse & Selve BuS.IVa engines and fitted with a lengthened fuselage and a box tail to improve engine-out control was flight tested with good results. The box tail had two rudders and biplane horizontal stabilizers and elevators to give the aircraft better controllability during engine-out operations, and succeeded to the extent that the aircraft could even be turned toward the running engine. The modified aircraft, designated AEG G.IVb-lang, (*lang* = long, for the lengthened fuselage) was the forerunner of the AEG G.V.

On 30 July 1919, AEG test pilot Paul Schwandt and seven passengers (1,000 kg useful load) broke the official world record by reaching 6,100 meters in the AEG G.IVb-lang 856/17. At the time, the record machine was powered by two 260 hp Mercedes D.IVa engines supercharged by two AEG turbo-compressors driven by shafts from the rear of the engines.

Above & Below: AEG G.IV G.1125/16 from the second G.IV production batch is shown here in French markings after being brought down on December 23, 1917 by anti-aircraft fire. It wears AEG-style lozenge night-bomber camouflage.

Above: AEG G.IV G.581/17 from the third G.IV production batch wore the AEG lozenge night-bomber camouflage and intermediate-style thick version of the straight-sided *Balkenkreuz* national insignia.

Below: AEG G.IV G.1131/16 from the second G.IV production batch is being evaluated with other captured German aircraft, including an LVG C.V and Albatros D.Va, at the French aviation test center at Villacoublay.

Above: AEG G.IV G.1131/16, slightly bent after its hard forced landing at night, at Villacoublay shows details of its engine installation, camouflage, and bomb racks. Storage for signal flares is provided on the outside of the front gun position.

Below: AEG G.IV G.1256/18 from the last G.IV production batch wearing late-style national insignia. It was turned over to the British in late December 1918 in accordance with Armistice requirements.

AEG G.IV

AEG G.IV prototype.

AEG G.IV G.155/16, the 3rd G.IV built.

AEG G.IV G.572/17 brought down behind American lines on 2 June 1918.

Right & Below: More images of AEG G.IV G.1256/18 from the last G.IV production batch. These were taken postwar and the German national insignia have been painted over by the new owners. The photos were taken at Bickendorf airfield near Cologne (Köln).

Below: AEG G.IV closeup showing the landing lights in the lower nose and the numerous bombs mounted under the wings and fuselage. The finish is the typical AEG lozenge night camouflage.

Above: AEG G.IV with a ferocious face and heavy bomb load wears late-war insignia and "VII" on the fuselage side.
Below: AEG G.IV G1125/16 was given captured aircraft number G.105.

Above: This front view shows how compact the AEG G.IV design was.

Above: AEG G.IV of *KG*4 in 1918.

Right & Below Right: The national insignia of these AEG G.IV bombers were cut out after capture.

Below: This AEG G.IV was repainted in French markings after capture. The Roman numeral 'III' may indicate it was assigned to *Bogohl* 3.

Above: AEG G.IVb G.189/16 was one of the first production batch rebuilt with extended, three-bay wings to carry heavier bomb loads. The reason for the larger wing, a 1,000 kg P.u.W. bomb, is being loaded.

Below: AEG G.IV surrounded by British troops after capture.

AEG G.IV

AEG G.IV G.567/18 of *Bogohl* VIII, *Staffel* 27. The "27" on the fin indicates *Staffel* 27; "7" is its tactical number within the *Staffel*. Normally there were six aircraft to a bombing *Staffel* so this was unusual.

AEG G.IV G.848/17, *Bogohl* VIII, *Staffel* 27, perhaps the aircraft of the *Staffel commander*, *Oblt*. Fritz Diemer, Maria Alter Aerodrome, May 1918.

AEG G.IVb G.168/16.

AEG G.V

In November 1917 AEG began work on a night bomber designed to carry a useful load of 2,100 kg, including a 1,000 kg P.u.W. bomb. The new bomber was based on the earlier AEG G.IVb-lang but with wingspan and area further increased for greater payload. The engines reverted to the 260 hp Mercedes D.IVa that powered production AEG G.IV bombers. The longer fuselage and box tail of the AEG G.IVb-lang were retained for improved engine-out handling.

To further improve flying qualities, the G.V used the Flettner *Hilfsruder* (servo rudder) that aerodynamically reduced the control forces and made large bombing aircraft less tiring to fly. Developed as part of Anton Flettner's work on automatic guided-missile control, the Flettner servo rudder was an advanced design feature of all late-1918 German bombers. According to one historian "the British Air Ministry technical reports of 1919 indicate complete ignorance of the aerodynamic principles employed!"

In May 1918 the AEG G.V prototype reached 4,000 meters in 70 minutes carrying a useful load of 2,100 kg. *Idflieg* held a competition between the AEG G.IV, AEG G.V, Friedrichshafen G.IV, and Gotha G.Vb bombers and found that, when carrying a useful load of 1,600 kg, the AEG G.V and Friedrichshafen G.IV were comparable in performance, and both were superior to the AEG G.IV and Gotha G.Vb in speed and climb. The first order for the G.V was placed in April 1918 and the first nine production G.V bombers were delivered in August. Between August and October 1918 a total

Above & Below: Similar views of AEG G.V 301/18 showing its enlarged, three-bay wing with Flettner tabs to reduce the pilot's aileron control forces. The 'box' tail with twin rudders gave better engine-out handling.

Above: AEG G.V anchors a display of late-war aircraft, perhaps postwar.

Above: The Fokker V17 monoplane fighter prototype provides an interesting contrast to the AEG G.V and emphasizes the size of the G.V.

Above & Below: On November 29, 1918, less than three weeks after the end of the war, AEG sold six AEG G.V bombers to Sweden. In fact, the first G.V, 1708/18, had already arrived at Eksjö on 23 November!

of 37 G.V bombers were delivered from the 50 that were ordered.

Postwar, AEG sold six AEG G.V bombers to Sweden on November 29, 1918, with AEG to deliver the aircraft by air. The first G.V, 1708/18, had already arrived at Eksjö on 23 November 1918. Three G.V bombers left Hennigsdorf on 29 March 1919 but two force landed before reaching Sweden and the third was destroyed on take-off in Vilemölla on 4 April 1919. In May–June 1919 two additional G.V bombers (including G.1712/18) were flown to Kristianstad, followed by two further replacements (including G.1710/18) in August and September 1919.

Above & Below: Additional views of AEG G.V 644/18 showing more detail of its ailerons with Flettner tabs to reduce the pilot's aileron control forces. The AEG G.V was the best two-engine German bomber in service at war's end.

Above: Front view of AEG G.V 644/18 emphasizing the three-bay wings and distinctive Flettner tabs on the ailerons.

Below: Side view of an AEG G.V 644/18; power was from a pair of 260 hp Mercedes D.IVb engines like the earlier G.IV.

Above: Side view of an AEG G.V of the second batch showing the Flettner tabs and 'box' tail with twin rudders.
Below: Rear view of an AEG G.V showing the 'box' tail with twin rudders to better advantage.

Albatros Bombers

The Albatros company was the largest German aircraft manufacturers and was famous for its fighters and two-seat reconnaissance airplanes. In addition to these types, Albatros also designed and manufactured several G-type bombers, of which one, the G.III, saw service in limited numbers.

Albatros G.I

Above & Below: A single Albatros G.I was built but the type was not satisfactory.

Ostdeutsche Albatroswerke G.m.b.H (OAW), part of the Albatroswerke G.m.b.H, was one of two Albatros companies until they merged in October 1917. Chief engineer at OAW was *Dipl.-Ing.* Karl Grohmann, who designed the Albatros G.I. Only a single Albatros G.I was built at OAW's Schneidemühl factory. Interestingly, it was a rare, four-engine aircraft with three bays of bracing outboard of the tractor engines. The G.I was powered by four 120 hp Mercedes D.II engines and made its first flight on January 31, 1916, but no further details are available. For unknown reasons the aircraft was unsatisfactory and no further development was undertaken.

Albatros Bomber Specifications

	G.I	G.II	G.III
Engines	4 x 120 hp Mercedes D.II	2 x 150 hp Benz Bz.III	2 x 220 hp Benz Bz.IV
Span (Upper)	30 m	17.0 m	18.0 m
Span (Lower)	—	17.0 m	17.0 m
Length	—	11.9 m	11.9 m
Height	—	4.2 m	4.2 m
Empty Weight	—	—	2,004 kg
Loaded Weight	4,319 kg	—	3,086 kg
Max. Speed	—	—	150 km/h
Climb, 1,000 m	—	9.3 minutes	10.0 minutes
Climb, 2,000 m	—	25 minutes	30 minutes
Climb, 3,000 m	—	70 minutes	45 minutes
Service Ceiling	—	3,000 m	3,000 m
Range/Endurance	—	4 hours	595 km
Bomb Load	—	—	325 kg
Armament	—	2 flexible MGs	2 flexible MGs

Note: Only one G.I and one G.II were built. The highest number of Albatros bombers at the front was 9, indicating a short production run. Known serials were in the range G.126–132/16.

Albatros G.II

Above: The sole Albatros G.II prototype was a compact bomber with thick airfoil, equal span wings. The engines were mounted as pushers above the lower wing in streamlined nacelles. The single-bay interplane struts were very robust and braced to maintain rigging with minimum assistance from the bracing wires. The rudder is aerodynamically balanced and a nose landing gear was fitted to prevent nose-overs on landing.

At the main Albatros factory at Johannisthal, Robert Thelen, assisted by George Madelung, designed the two-engine Albatros G.II based on the *Kampfflugzeug* specificatins promulgated by *Idflieg* in July 1914. Madelung was responsible for the thick airfoil section chosen for the G.II that was intended to give it greater lift than the thin sections then common. The prototype C.IV was built specifically to test this thick airfoil section for the G.II.

The G.II fuselage was covered with plywood like other Albatros designs of that time and carried three crewmen, a pilot in the middle and fore and aft gunners. The massive interplane struts eliminated the need for incidence bracing wires and the landing gear resembled that of a two-seater except for the nose gear. With 150 hp Benz Bz.III engines the aircraft was reliable but under-powered, and interest moved on to a more powerful derivative, the G.III.

Left: The Albatros G.II was a compact, streamlined bomber with good flying characteristics. However, climb and ceiling were poor and it was evident that more power was needed than the 150 hp Benz Bz.III engines delivered. Design of the G.II was unrelated to the G.I other than company name.

Below: The sole Albatros G.II looked like a member of the Albatros family of aircraft with the exceptions of the thick airfoil and the balanced rudder needed to give more directional control in case of engine failure.

Above: The sole Albatros G.II prototype under test at Johannisthal. Its thick airfoil, equal span wings, and distinctive single-bay interplane struts were hallmarks of this interesting design.

Albatros G.III

Above: This appears to be the Albatros G.III prototype. The engines are mounted above the lower wing in streamlined nacelles, the single-bay interplane struts are normal size, and the upper wing span has been increased but the ailerons have no aerodynamic balances. The G.III discarded the nose landing gear used by the G.II.

The Albatros G.III was based on the G.II and used a similar, perhaps identical, plywood-covered fuselage. More powerful engines, 220 hp Benz Bz.IV inline six-cylinder types, were fitted to improve climb and bombload. The thick airfoil section was retained, and to further improve load-carrying capability the upper wing span was increased by a meter for more lifting area. At first the ailerons were similar to those of the G.II, but later aerodynamic balances were fitted to reduced the pilot's control forces and improve maneuverability. However, the aerodynamic balance on the rudder was deleted, apparently to harmonize the control forces.

The landing gear was modified to eliminate the nose gear and dual wheels were fitted to each main gear. More conventional interplane bracing was used, although the G.III retained the single-bay configuration. Installation of the more powerful engines was similar to the G.II but the mounting was more streamlined. However, cooling problems must have afflicted the G.III because nearly all photos of operational G.III bombers show them without any cowling panels for improved cooling at the expense of streamlining.

Despite the improvements compared to the G.II, the G.III still had a lighter bomb load than its rivals and its flying and landing characteristics were not as good as desired. Regardless, *Idflieg* ordered a small number of G.III bombers, perhaps 10–12, for operational evaluation. In early 1917 some were assigned to *Kagohl* 4 in the Balkans, and others were assigned to *Kagohl* 2 on the Western Front. The G.III front-line inventory peaked at nine aircraft in April and only one was at the front by the end of 1917.

The Albatros G.III thus made a limited contribution to the German war effort and was the least successful operational bomber. Its qualities were such that Albatros abandoned further development of two-engine bombers to focus on their more successful fighters and two-seaters.

Above: This rear quarter view of the Albatros G.III prototype emphasizes its aerodynamic design for a two-engine bomber of the period. The engines are mounted in streamlined nacelles, the wing bracing is single-bay, and the landing gear is simple. The tail surfaces are shaped like the successful, contemporary Albatros two-seaters and fighters, although the fuselage was rectangular in cross section, unlike Albatros fighters with their oval fuselage designs. The G.III had three crewmen, a pilot and fore and aft gunners, and each gunner had a single, flexible machine gun. The thick airfoil section was chosen for high lift and within Albatros designs was unique to the G.II, G.III, and the C.IV airfoil testbed. Flight testing led to addition of aerodynamic balances on the ailerons to reduce control forces and improve maneuverability. The number of Albatros G.III bombers built is not known; but the maximum front-line inventory was nine in April 1917.

Above & Below: Albatros G.III G.126/16 serving with an operational unit. The thick airfoil section is prominent as are the bomb racks. Although the spinners have been retained, the engine cowling has been removed for additional cooling. The wood grain of the ply-covered fuselage is noticeable in the photo above, but appears to have been painted in the photo below, probably for duty as a night bomber because the paint appears as dark as the national insignia under the wing, but lighter on the fuselage, probably as a result of the sunlight.

Left & Below: Albatros G.III G.130/16 *Hansi* at an operational unit. The thick airfoil section and aerodynamically-balanced ailerons are unlike other contemporary Albatros warplanes, but the shape of the tail surfaces is very similar to that of contemporary Albatros fighters and two-seaters.

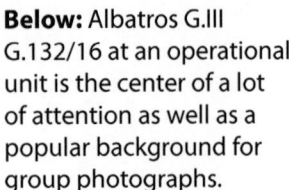

Below: Albatros G.III G.132/16 at an operational unit is the center of a lot of attention as well as a popular background for group photographs.

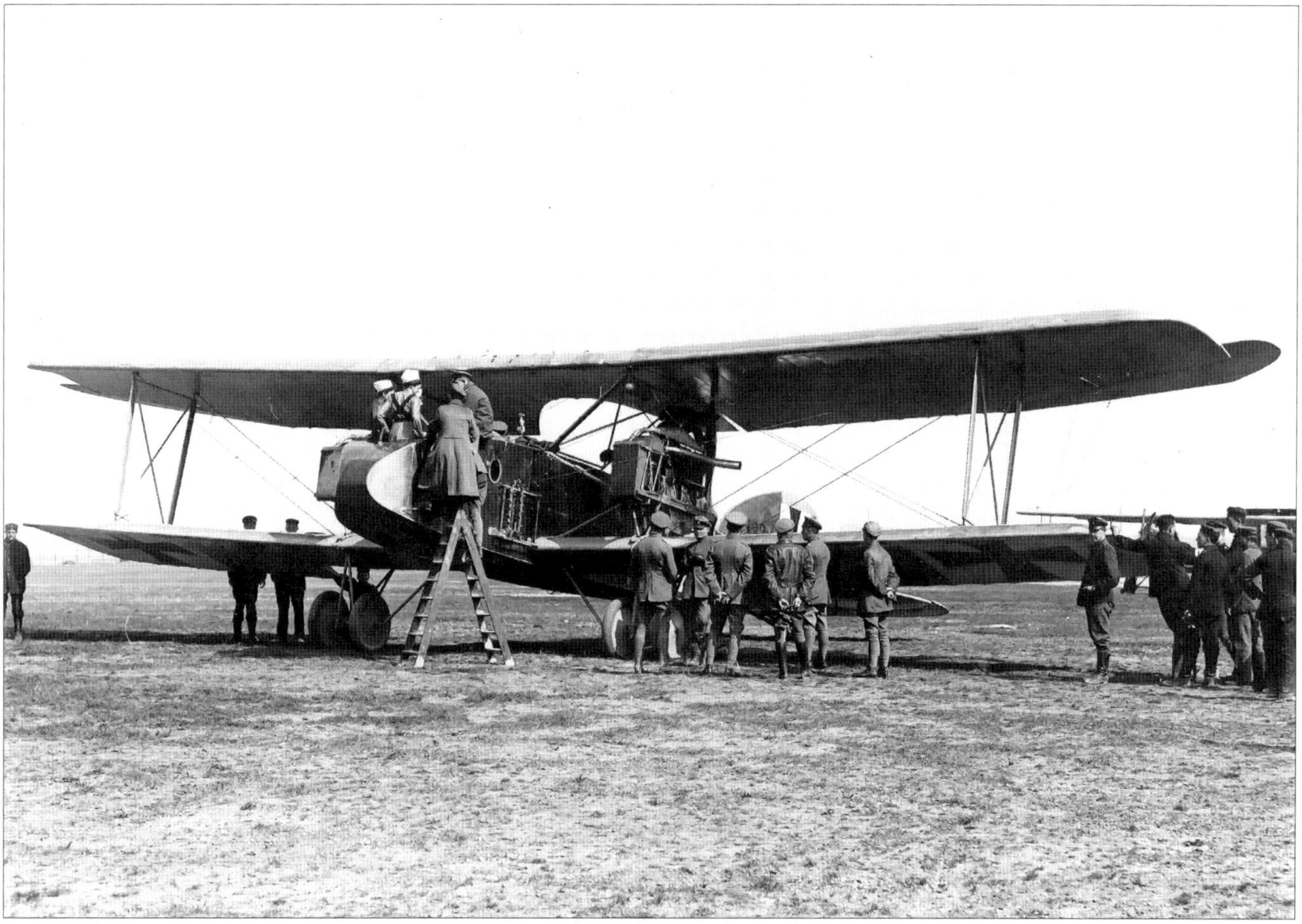

Above & Below: Albatros G.III G.132/16 at an operational unit. The thick airfoil section – unlike other contemporary Albatros warplanes – and fuselage bomb racks are prominent. The spinners are missing and the engine cowlings have been removed for improved cooling despite the increased drag produced.

Right & Facing Page: Albatros G.III in dark colors displays its bomb load. Six heavier bombs (in this case apparently 50 kg *P.u.W.* bombs) are carried under the fuselage near the center of gravity and smaller (12.5 kg *P.u.W.*) bombs are carried in racks on the fuselage sides. The bomb load could total at least 325 kg, and this G.III appears to be carrying more than that.

Above & Below: Landing accidents with two different Albatros G.III bombers had similar results. The aircraft above was flying in summer and the engine cowlings had been removed, while the aircraft below, likely the prototype, was flying in winter and had its engine cowlings fitted.

Left: Aircrew and others relax around their Albatros G.III bomber with its engine cowlings removed. That nearly all photos show the G.III without engine cowlings indicates cooling problems.

Below: Aircrew and visitors relax around an Albatros G.III bomber in the field. Like most photos showing operational G.III bombers, it has had its engine cowlings removed.

Below: This Albatros G.III bomber had its engine cowlings removed even when flying in winter.

Albatros G.III

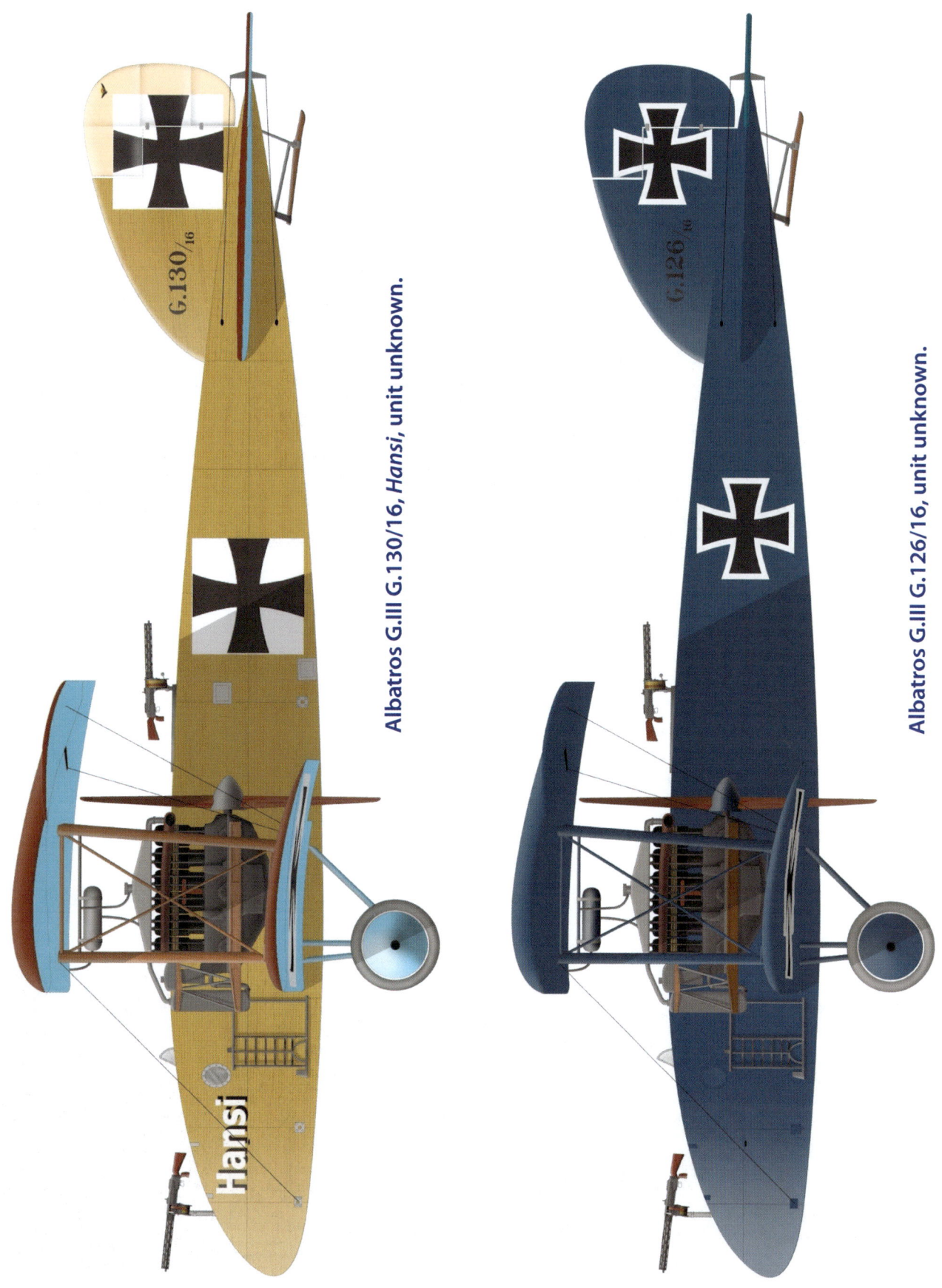

Albatros G.III G.130/16, *Hansi*, unit unknown.

Albatros G.III G.126/16, unit unknown.

SSW L.I & LVG G.III

LVG G.III prototype, late 1918.

SSW L.I prototype, late 1918.

Daimler Bombers

During WWI the Daimler company is best known for its Mercedes engines that powered so many German warplanes. Much less known is that Daimler ran a major aircraft repair facility, built 298 Friedrichshafen bombers under license, and eventually designed and built its own aircraft, including aircraft in the G, R, D, and CL categories, none of which, for various reasons, reached production status. The Daimler bomber story is related below.

Union G.I (R.I)

On August 6, 1915, the *Feldflugchef* (Chief of Field Aviation) suggested pressing Daimler to manufacture multi-engine G-type aircraft, and Daimler agreed to expand into aircraft production.

Daimler's first bomber involvement was with the Union G.I (also known as the R.I), where it provided the engines. The Union G.I was a biplane with four 110 hp Mercedes engines of inverted configuration. The engines were mounted as tractor-pusher combinations in nacelles attached to the lower wing. Four engines were installed for improved reliability, and the tractor-pusher combination enabled the designers to mount all the engines near the fuselage to minimize asymmetric thrust due to engine failure. The compact design also provided good fields of fire to the fore and aft gunners, each of whom were to have a single flexible machine gun.

Below: The Union G.I had four 110 hp Mercedes engines mounted in tandem pairs.

On May 17, 1915 Daimler agreed to inspect the aircraft, and especially the engines, prior to its first flight that was scheduled for the following week. At some point during its flight testing the Union G.I carried 18 people, an unofficial world record, before being accepted by the *Fliegertruppe*. The Union G.I was transferred to *Feld-Flieger Abteilung* 26 and was attached to that unit from June 1 to its crash on September 1 at Schloss Eberspark near Schneidemühl.

By the time of its crash the Union G.I had been named *Marga-Emmy*. With *Lt.* Neuensitz as observer, the pilot Thässler landed the G.I hard, breaking off the tail. Although the nose dug in, neither crewmen was injured. However, that was the end of the Union G.I. More importantly, it was the end of the Union company; the *Fliegertruppe* cancelled any further support and the company had to sell its assets to Norddeutsche Flugzeugwerke.

This Page: Three photos of the Union G.I. Daimler supplied the four 110 hp Mercedes engines mounted in tandem pairs. These engines were unusual in being inverted, that is, the cylinders were below the crankcase. The Union G.I was also known as the R.I (the "R" standing for *Riesenflugzeug* – Giant airplane) due to its four engines, but it was not a proper R-type because the engines were not accessible in flight as specified by the R-type requirements, so it was re-designated a G-type.

Above: The Union G.I was named *Marga-Emmy* after it was assigned to *Feld-Flieger Abteilung* 26.
Below: The Union G.I being readied for flight while at *Feld-Flieger Abteilung* 26.

Above: The Union G.I taking off.
Below: The Union G.I experiencing a take-off accident.

Above: The tail of the Union G.I is carried back to the hangars after its landing accident on September 1, 1915 at Schloss Eberspark (Schneidemühl). The pilot, Thässler, made a hard landing and broke the tail off the aircraft. The G.I was not repaired after this accident.

Above: The end of the Union G.I and the Union company after its landing accident on September 1, 1915 at Schloss Eberspark (Schneidemühl). The pilot, Thässler, and observer *Lt.* Neuensitz were unhurt and pose with the aircraft. The tail has broken off the fuselage.

Daimler R.II

After the demise of the Union G.I and the Union company, Daimler got more directly involved in building bombers by hiring the designer of the Union G.I, *Ingenieur* Karl Schopper, who designed two similar aircraft for Daimler.

Daimler had already received orders from *Idflieg* for two bombers that apparently were not built. The first was for a two-engine bomber powered by a pair of 220 hp Mercedes D.IV straight-eight engines. *Idflieg* also ordered six G-types with serials G.112–117/16, but there were apparently not built either as no records or photographs of these aircraft have been found.

Apparently the first Daimler bomber that was actually built was the Daimler R.II 450-1915, a design that was clearly a follow-on to the Union G.I. The designation 'R.II' may have been chosen to follow the original 'R.I' designation of the Union bomber, but neither was a true R-type because the engines could not be accessed and worked on in flight as required by the R-type specifications.

In any case, the Daimler R.II was powered by four 100/110 hp Mercedes engines and was otherwise very similar to the earlier Union G.I by the same designer. The main visible difference was the engines of the Daimler R.II were of conventional configuration with the cylinders above the crankcase. This placed the propellers much lower relative to the engine nacelles.

The R.450 flew in November 1916 and was active until April 1917. Flight testing did not go well, with a series of engine problems, structural problems, and poor flying qualities. The aircraft was very difficult even to get into the air and when in flight was not rigged properly so had a tendency to turn to the right. After an unsuccessful attempt to take off on April 13, 1917, there are no more records discussing flights and flight testing was apparently abandoned.

Above & Below: The Daimler R.II was designed by Karl Schopper, who had designed the Union G.I. The key visible difference between the types was the location of the propellers, which was dictated by the type of Mercedes engine used.

Above: The Daimler R.II was structurally weak and had very poor flying qualities. Development was abandoned after test pilot Irrek was unable to get the aircraft into the air for another test flight after five attempts.

Daimler R.I

Confusingly, the Daimler R.I 478-1915 was built after the R.I 450-1915, which is why it is presented here out of numerical order. The Daimler R.I was another variation on the Union G.I/Daimler R.II theme. Like the earlier R.II the R.I was powered by four 100/110 hp Mercedes engines. The only visible difference was in the engine nacelle design, but there were a number of internal structural improvements that were not externally visible.

Test pilot Irrek, accompanied by second pilot Gaiser, probably made the first flight of the R.I on January 18, 1917. Like the preceding R.II the R.I was not rigged correctly and tended to bank right and turn right. Speed was 120–125 km/hr but the fore and aft engines did not run at the same RPM, and the controls were very stiff. A second test flight on January 22 was little better. The engines now ran at about the same RPM, so the propellers

Below: The hapless Daimler R.I was little improved over the unsatisfactory R.II; development was abandoned in May 1917. Both flying qualities and performance were very mediocre.

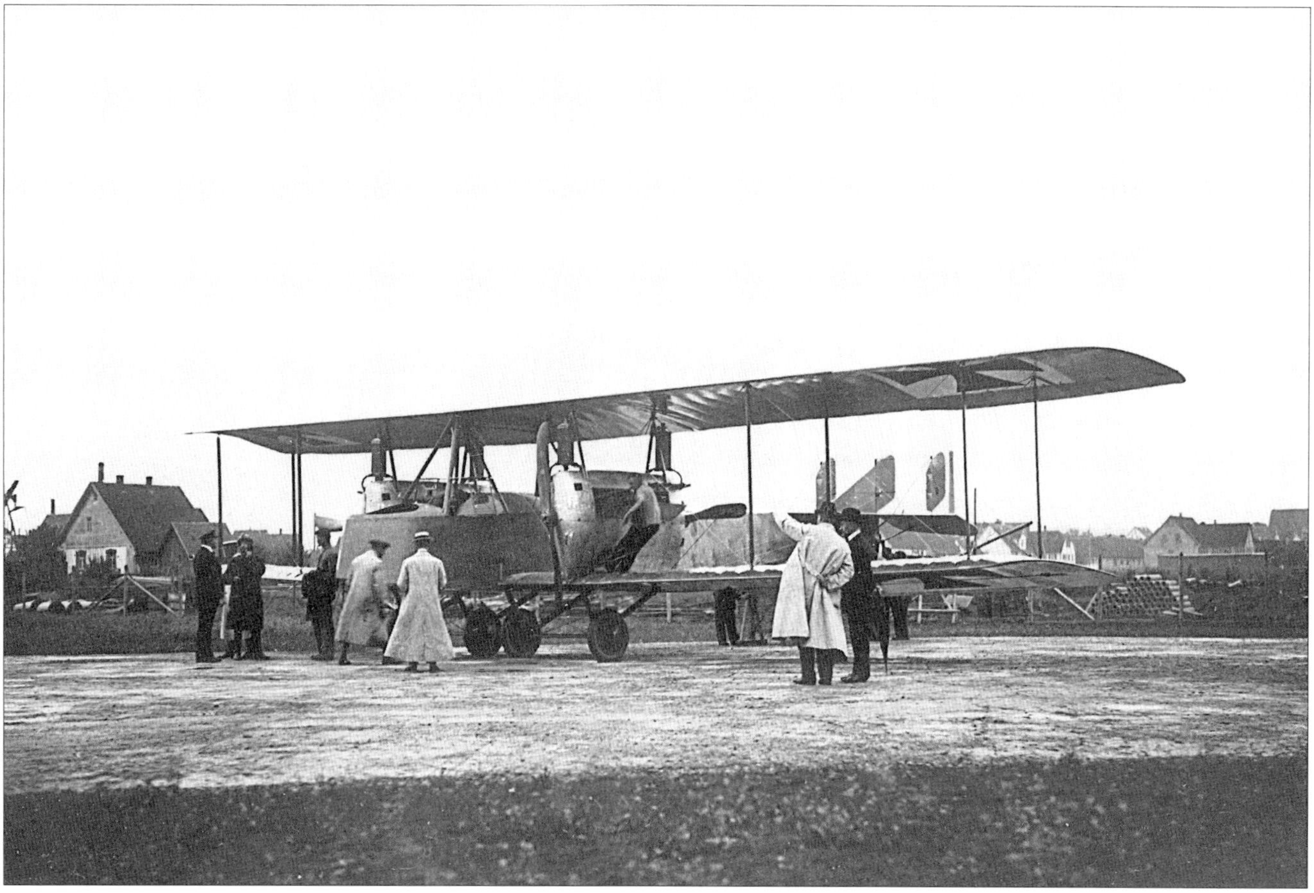

Above & Below: The Daimler R.I was can be distinguished from the earlier R.II by its re-designed engine nacelles. Internal improvements were also made but were not visible. Poor performance and flying qualities quickly doomed the R.I.

had apparently been changed. However, it took 62 minutes for the R.I to reach 2,400 meters, a very mediocre performance.

Irrek made the next test flight on April 26 solo and could barely reach 500 meters altitude. The flight was subject to intense engine vibration and Irrek was concerned about fuselage structural failure. On May 24 Irrek and Gaiser made an acceptance flight of 1.25 hours duration and reached 2,900 meters.

The May 24 flight was probably the last for the R.I or any other four-engine aircraft designed by Schopper. These aircraft had proved complex, underpowered, and were not robust. Both flying qualities and performance were mediocre at best, and the pilots did not regret their passing.

Daimler G.I

Daimler's first two-engine bomber design, the G.I, ordered as 476-1915, was not completed and no further information about it is available.

Below: The Daimler G.I being built in the Daimler factory; it was not completed and no details are known.

Daimler G.II (480-1915 Tractor)

The Daimler G.II was designed by Schopper but was a completely different aircraft than the earlier Union G.I and the related Daimler R.I and R.II. The Daimler G.II, aircraft 480-1915, was a two-engine bomber with its 220 hp Mercedes D.IV engines mounted as tractors.

Daimler test pilot Irrek flew the G.II (tractor) on November 20, 1916, which may have been its first flight. Unfortunately, like the earlier R.I and R.II the G.II 'hung right' (banked to the right) and turned right and the ailerons were ineffective because the upper wing bent under the aerodynamic loads and stretched the aileron control cables. The rudder control was too heavy and the elevators were light. The aircraft was tail heavy and Irrek could not maintain directional control with one engine throttled back to simulate engine failure. Although six aircraft had been ordered, development of the tractor version of the G.II may have been stopped at this point because there is no more information about further tests flights, additional airframes, or technical modifications.

Daimler G.II (Tractor) Specifications		
Engines:	2 x 220 hp Mercedes D.IV	
Wing:	Span Upper	17.0 m
General:	Loaded Weight	2,700 kg

Above & Below: The Daimler G.II 480-1915 (tractor) in the factory during assembly. Workmanship was excellent.

Above: The Daimler G.II 480-1915 (tractor) in the factory during construction. The Daimler R.I or R.II is in the foreground.

Below: The Daimler G.II 480-1915 (tractor) in the factory after completion. The lower wing was significantly smaller than the upper wing and test pilot Irrek thought this contributed to the upper wing bending experienced in flight.

Daimler G.II (480-1915 Pusher)

Schopper also designed a pusher version of the Daimler G.II that was very similar to the tractor version but had a larger lower wing, now of equal span with the upper wing. Daimler test pilot Irrek made seven unsuccessful take-off attempts in this aircraft between March 16 and April 19, 1917. On June 28 Irrek was finally able to get the recalcitrant G.II (pusher) into the air, but its flying qualities were very poor. Irrek was barely able to land the aircraft safely after a nerve-wracking flight. To the relief of the pilots, further development of the Daimler G.II was abandoned.

Above & Below: The Daimler G.II 480-1915 (pusher) was powered by the 220 hp Mercedes D.IV straight-eight like the preceding G.II (tractor). Also like the G.II (tractor), workmanship was excellent but the design had dangerous flying qualities.

Above, Below, & Bottom: The Daimler G.II 480-1915 (pusher) was well-built but unfortunately was not well designed. Its excellent workmanship was wasted due to dangerous flying qualities and the design was quickly abandoned.

Daimler G.III (584-1916)

The Daimler G.III was powered by two 260 hp Mercedes D.IVa engines buried in the fuselage. Once again Schopper was the designer, a fact difficult to understand in view of his consistent record of designing bombers with poor performance and poor to dangerous flying qualities. The internal engines were coupled together and drove two tractor propellers via drive shafts and gears. The vertical tail was similar to that of the G.II.

As originally built, the bulky G.III had a tricycle landing gear and a huge block radiator mounted above the nose to facilitate flight trials. Taxi trials were initiated on July 17, 1917, and the next day test pilot Irrek made the first flight. A number of test flights followed, many of short duration which indicated problems. After a flight on September 8 the G.III was grounded for more than a month for either repairs or modifications. A duration flight was made on October 16 with full load and a crew of three; cruise speed was 120 km/h but the aircraft was very tail-heavy, making longitudinal stability marginal. Irrek thought the relatively short fuselage contributed to inadequate directional stability as well. Different wings were to be fitted to improve flying qualities, but it is not clear the aircraft was flown after this major modification. Daimler was too busy with building and testing Friedrichshafen bombers it was building under license to invest more resources on the G.III, which was regarded as too heavy due to the central drive system. Daimler G.III development was stopped, leaving its designer Schopper with a record unblemished by success.

Above & Below: As built the Daimler G.III 584-1916 featured a nose landing gear and a massive block radiator on the nose to expedite flight testing. Two 260 hp Mercedes D.IVa engines were buried in the fuselage to provide power.

Left: The shape of the Daimler G.III vertical tail surfaces was very similar to the earlier G.II. It is surprising that Daimler continued to have Schopper design its bombers given that none of his designs were successful. All of his bomber designs had poor flying qualities at best, and were dangerous at worst.

Above: The Daimler G.III after the nose gear was removed (the stubs remain visible) and the block radiator was replaced by smaller radiators seen either side of the crewman.

Left: The Daimler G.III after the nose gear was removed and the block radiator was replaced.

Above: All Daimler bombers were well built but poorly designed. The central power system of the G.III was too heavy and flying qualities were poor.

Right: The Daimler G.III was a bulky design due to the internally-mounted engines.

Below: The Daimler G.III during flight testing showing the power transmission system and propeller-bracing details.

Friedrichshafen Bombers

Although the Friedrichshafen bombers are not nearly as well known as the famous Gotha that bombed London in daylight and later at night, they were actually superior aircraft that served competently as tactical bombers, primarily attacking Allied targets behind the lines at night.

The Friedrichshafen bombers were sturdy, reliable, and perhaps most important to their crews, had good handling qualities that made them much easier to land than the Gothas, which reduced the frequency of their landing accidents with resulting casualties.

Friedrichshafen G.I

Only a single Friedrichshafen G.I, 118/15, was built in 1915. This prototype was powered by two 150 hp Benz Bz.III engines mounted as pushers, and it featured three-bay wing bracing and a biplane tail assembly. Built to the original *Kampfflugzeug* (battle plane) specifications that *Idflieg* issued in 1914, the *Kampfflugzeug* concept was basically an 'aerial cruiser' armed with machine guns and bombs. The aircraft was to have 200 hp, carry a crew of three, and have an endurance of six hours. Unfortunately, the battleplane was a flawed concept.

The Friedrichshafen G.I, factory designation FF36, had a crew of three and a single flexible machine gun in the front gunner's cockpit. Construction of the G.I was the conventional wood frame with wire bracing and fabric covering common for the time. Testing was sufficiently promising that *Idflieg* ordered an improved follow-on design, the G.II.

Below & Facing Page: The Friedrichshafen G.I set the basic configuration for all following Friedrichshafen bombers. Only one prototype was built and evaluated before *Idflieg* ordered an improved, more powerful derivative, the Friedrichshafen G.II. (Peter M. Bowers Collection, Museum of Flight)

Friedrichshafen Bomber Orders				
Date	Type	Qty	Serials	Notes
1915	G.I	1	G.118/15	
Late 1915	G.II	6	G.131–136/16	
Feb. 1916	G.II	11	G.100–113/16	G.104/16 was the G.III prototype
Oct. 1916	G.II	12	G.625–636/16	Daimler built. G.637–648/16 not built.
Feb. 1917	G.II	6	G.150–155/17	Daimler built
Nov. 1916	G.III	24	G.1030–1053/16	
May 1917	G.III	25	G.156–180/17	
May 1917	G.III	30	G.240–269/17	Daimler built
June 1917	G.III	100	G.270–369/17	
July 1917	G.III	60	G.373–432/17	Daimler built
Sep. 1917	G.III	6	G.554–559/17	Daimler built
Dec. 1917	G.III / IIIa	60	G.784–843/17	
Dec. 1917	G.III	4	?	Daimler built; serials in range G.433–442/17?
Feb. 1918	G.III	100	G.355–414/18 G.505–544/18	
Feb. 1918	G.III	35	G.150–184/18	Hanseatische built; serials unconfirmed
April 1918	G.III	100	G.645–744/18	
April 1918	G.III	50	G.1400–1449/18	
April 1918	G.III	50	G.1793–1827/18	Ordered from Gotha; not built
April 1918	G.III / IIIa	30	G.750–779/18	Daimler built
June 1918	G.IIIa	50	G.780–829/18	
June 1918	G.IIIa	30	G.880–909/18	Daimler built
July 1918	G.IIIa	150	G.830–879/18 G.1160–1209/18	
July 1918	G.IIIa	45	G.1000–1044/18	Daimler built
July 1918	G.IIIa	40	G.1120–1159/18	Hanseatische built
Sep. 1918	G.IIIb / G.IV	100	G.1465–1564/18	
Oct. 1918	G.IIIa / G.IV	75	G.1045–1119/18	Daimler built
1918	G.V	3	G.900–902/17?	FF62 used the G.V airframe, Mercedes engines

Above: The Friedrichshafen G.I during flight evaluation, which revealed the need for more power. (Peter M. Bowers Collection, Museum of Flight)

Below: The completed Friedrichshafen G.I being rolled out of the factory. (Peter M. Bowers Collection, Museum of Flight)

Friedrichshafen Bomber Specifications							
	G.I	**G.II**	**G.III**	**G.IIIa**	**G.IIIa(Daim)**	**G.IV**	**G.V**
Engines	2x150 hp Benz Bz.III	2x200 hp Benz Bz.IV	2x260 hp Mercedes D.IVa	2x260 hp Mercedes D.IVa	2x260 hp Mercedes D.IVa	2x260 hp Mercedes D.IVa	2x245 hp Maybach Mb.IVa
Span	—	19.75 m	24.0 m	24.0 m	24.05 m	22.60 m	19.20 m
Wing Area	—	75.5 m²	94.26 m²	94.86 m²	—	87.00 m²	76.00 m²
Gap	—	2.10 m	2.11 m	—	—	—	—
Length	—	11.05 m	12.78 m	12.24 m	12.65 m	12.00 m	10.00 m
Height	—	3.41 m	3.54 m	—	3.40 m	—	—
Empty Wt.	—	2,180 kg	2,560 kg	2,700 kg	3,000 kg	2,880 kg	2,410 kg
Loaded Wt.	—	3,180 kg	3,795 kg	4,200 kg	4,500 kg	4,980 kg	4,214 kg
Useful Load	—	1,000 kg	1,235 kg	1,500 kg	1,500 kg	2,100 kg	1,804 kg
Max Speed	—	150 km/h	150 km/h	140 km/h	140 km/h	142 km/h	—
Climb: 1,000 m	—	—	3.5 min.	—	—	—	—
2,000 m	—	—	10.5 min.	—	—	—	—
3,000 m	—	—	20 min.	—	—	—	—
4,000 m	—	—	36 min.	—	—	—	—
5,000 m	—	—	74.5 min.	—	—	—	—

Notes: 1. The G.I & G.V were fitted with one flexible machine gun; the other types had two flexible machine guns.
2. The G.V airframe was also tested with 260 hp Mercedes D.IVa engines.

Below: The Friedrichshafen G.I under construction. When more powerful engines were added to later Friedrichshafen designs, larger, taller rudders were required. (Peter M. Bowers Collection, Museum of Flight)

Above: The Friedrichshafen G.I during construction; the G.I was designed without a rear gunner's position, a feature added to all operational Friedrichshafen bombers. Other than that, the subsequent Friedrichshafen bombers designs generally followed the basic configuration and structure of the G.I. (Peter M. Bowers Collection, Museum of Flight)

Friedrichshafen G.II

After evaluating the Friedrichshafen G.I, *Idflieg* ordered six examples of an improved, more powerful derivative, the Friedrichshafen G.II, factory designation FF38. Following the basic configuration and construction of the G.I, the G.II differed by having two 200 hp Benz Bz.IV engines, more compact two-bay wing bracing, and a conventional monoplane tail, although the first prototype was also tested with a 'box' tail for comparison. The specified bomb load was 150 kg, an indication the *Kampfflugzeug* was evolving into a bomber. Another indication was that a second flexible machine gun was installed for a rear gunner.

These first aircraft had serial numbers G.131–136/15, indicating they were ordered in late 1915. A second batch of a dozen G.II bombers, serials G.100–111/16, was ordered in February 1916. Unfortunately, the prototype was damaged in a crash during testing. Details are not known, but the result was the fuselage and lower wing structure had to be revised before the G.II was approved for production and operational service. The changes were made in May and the type testing was done in July and August, after which the modified prototype was approved for operational service.

However, the first six G.II bombers were apparently retained for a variety of test and experimental programs. For some reason additional modifications to the wings were required, and these were load-tested on October 14. On November 24 a revised tail structure was load tested. Aircraft G.105/16 was accepted on December 1, 1916, an indication that G.II production was completed early in 1917. In addition to the 18 G.II bombers ordered from Friedrichshafen, a like number were ordered from Daimler as the Friedrichshafen G.II(Daim). Only 35 G.II bombers were delivered as G.104/16 became the G.III prototype.

Above: A Friedrichshafen G.II in flight shows its two-bay wing bracing and typical Friedrichshafen configuration. The horizontal struts between the engine and fuselage below the upper longeron indicate either a prototype or a Daimler-built aircraft. (Peter M. Bowers Collection, Museum of Flight)

Below: Friedrichshafen G.II 131/15, the G.II prototype, in the factory. Here it has been modified from its original configuration by a taller fin and balanced rudder. (Peter M. Bowers Collection, Museum of Flight)

Above: A Friedrichshafen G.II(Daim), characterized by the horizontal struts between the engine and fuselage attaching below the upper longeron. This may be Daimler test pilot Irrek after the successful acceptance flight of the first G.II(Daim) 625.17 on March 3, 1917.

Below: The prototype Friedrichshafen G.II, G.131/15, at the factory. This photo shows the aircraft before modifications to the fuselage structure because the horizontal struts between the engine and fuselage are attached below the upper fuselage longeron. This feature was shared by Daimler-built aircraft but production Friedrichshafen-built aircraft had the struts attached to the upper longerons. (Peter M. Bowers Collection, Museum of Flight)

Above: Friedrichshafen G.II G.134/15 from the first G.II production batch.

Below: Another view of Friedrichshafen G.II G.134/15 showing the dramatic nose art that was applied. The horizontal engine struts attached to the upper fuselage longeron characterized production Friedrichshafen-built aircraft after structural modifications were applied to the fuselage. Only prototype Friedrichshafen-built aircraft had these struts mounted lower as seen on G.131/15 on the facing page. (Peter M. Bowers Collection, Museum of Flight)

Above: A Friedrichshafen G.II(Daim) in the drainage ditch on the Daimler airfield at Sindelfingen. The ditch was the source of many complaints from Daimler test pilots. (Peter M. Bowers Collection, Museum of Flight)

Below: An unfinished Friedrichshafen G.IV (note the four-bay wings and box tail) with an SSW fighter in the foreground. The G.IV was basically a G.IIIa with strengthened wings for greater bomb-carrying ability. (Peter M. Bowers Collection, Museum of Flight)

Friedrichshafen G.III / G.IIIa

The Friedrichshafen G.III, factory designation FF45, was an enlarged, more powerful development of the Friedrichshafen G.II. Following the basic configuration and construction of the G.II, the G.III differed by having two 260 hp Mercedes D.IVa engines and larger three-bay wings to enable it to carry a heavier bomb load. The G.III also added a nose wheel under the forward gunner's cockpit to prevent nose-overs on landing. This was important because most German bomber losses were due to bad landings, especially at night. The G.III was under development as early as July 1916, well before the G.II reached the front or was even approved for operational service. As noted above, airframe G.104/16 became the G.III prototype.

In November 1916 *Idflieg* ordered 24 production G.III bombers, serials G.1030–1053/16, with the first becoming the type aircraft, the production standard. This aircraft was flying by March 18, but was not sent to Adlershof for the type test until the designer, Kober, was satisfied with its performance and flying qualities. To improve maneuverability, ailerons were added to the lower wings. G.III 1030/16 was approved for operational service on April 17 providing it successfully performed the three-hour flight, passed the static load tests, and some minor installation and structural defects were fixed. These were easily corrected and the G.III was at the front by June 1917.

The G.III was also produced under license by Daimler as the G.III(Daim), Daimler already having produced the Friedrichshafen G.II under license. *Idflieg* ordered the first Friedrichshafen G.III(Daim) bombers in May 1917. On August 10 Daimler-built bomber G.240/17 passed the three-hour acceptance flight and bomber G.241/17 completed the *Typenprüfung* (Type Test) at Adlershof on August 22. Daimler was required to correct more than 30 minor installation defects and the G.III(Daim) was then approved for operational use. In addition to Daimler, the Friedrichshafen G.III was also ordered from the Caspar Hanseatische Flugzeugwerke, a total of 35 aircraft being produced as the Friedrichshafen G.III(Hansa). The Type Test of this license-built G.III was not completed until July 1918.

Although the Gotha bombers became far more famous due to bombing London by day and later by night, the Friedrichshafen G.III was a better aircraft. Used for tactical bombing, the Friedrichshafen G.III had better flying characteristics than the Gotha, which became dangerously tail-heavy after dropping its bombs. In addition to being more stable in flight, the Friedrichshafen G.III also had a more robust landing gear. Both characteristics became more important as bombers transitioned to night bombing due to improved Allied defenses.

Different engines were tested to improve high-altitude performance. Maybach Mb.IVa engines of

Above: The Friedrichshafen G.III, which followed the basic configuration of all Friedrichshafen bombers, was the main production type. Its configuration was very similar to the Gotha series but the Friedrichshafen had much better handling qualities that made it a safer airplane to fly, especially during landing.

260 hp were tried in one G.III, which improved its climb rate to 5,000 m in 42 minutes, a welcome improvement over the standard aircraft. Another G.III(Daim), 245/17, was tested with the over-compressed 260 hp Mercedes D.IVaü engine in January 1918. However, a combination of limited availability of the altitude engines and the switch to night bombing, which was done at lower altitudes, meant the altitude engines were reserved for the high-altitude reconnaissance aircraft that needed them. High reliability was far more important to night bombers than high-altitude performance.

In fact, for night bombing the first priority was reliability of the airframe and engines, followed by increased bomb load. By January 1918 *Kogenluft* wanted the payload increased to 1,000–1,200 kg. Next in priority were good take-off and landing qualities, good flying qualities, maneuverability, and the ability to maintain controlled flight despite an engine failure.

Friedrichshafen addressed these needs with modifications to the basic G.III airframe. A version of the G.III with biplane tail created the G.IIIa. Another derivative with increased wing area became the G.IV, and yet another derivative with tractor propellers became the G.V. Flettner assisted controls were used on the G.IIIa and G.IV, and another G.III was tested with mid-wing ailerons, but these did not go into production.

G.IIIa

Interestingly, the biplane or 'box' tail had been used on the first Friedrichshafen bomber, the sole G.I prototype. Now a similar tail was tested on the G.III in two versions, with and without a central fin supporting the upper stabilizer. The purpose was to improve directional control after an engine failure. The box tail (*Kastensteuer*) without fixed fin was chosen for production after comparative flight trials were completed in late March 1918. With box tail and a number of additional modifications to improve operational effectiveness, the bomber was designated Friedrichshafen G.IIIa.

At *Idflieg's* urging during a bomber conference that started April 3, 1918, both Friedrichshafen and Daimler agreed to install the box tail on G.III bombers already in production. The additional changes were made when the G.IIIa entered production, the first being delivered by Friedrichshafen in June 1918, from which time the G.IIIa began to replace the G.III. At this same bomber conference it was decided to modify the bomber to enable crewmen to easily move between the forward and aft gun positions. This change

Above: Friedrichshafen G.III 1031/16 of *Kampstaffel* 3.

resulted in the G.IIIb designation, which essentially included all the G.IIIa modifications together with a passageway on the starboard side. The drawings were completed in July and an unknown number of bombers were completed as the G.IIIb.

The G.IIIa proper embodied a number of modifications in addition to the box tail. A new fuel system with increased fuel capacity enabled flight durations up to six hours. A Flettner servo-tab (*Hilfsruder* – assisted rudder) system was fitted to the upper ailerons to reduce the pilot's control forces for easier flight and improved maneuverability. For greater bomb load the standard useful load was increased to 1,500 kg, and the maximum permissible load was increased to 2,100 kg. The landing gear was modified to increase clearance to mount larger bombs under the fuselage. Finally, a Gotha-style gun tunnel was installed in the fuselage to improve the rear gunner's downward field of fire, a change necessitated by increased British night fighter activity. Daimler also changed to G.IIIa(Daim) production as did the Caspar Hanseatische Flugzeugwerke, a total of 40 aircraft being ordered as the Friedrichshafen G.IIIa(Hansa) on July 13, 1918.

Austro-Hungarian Involvement

The Austro-Hungarian *Luftfahrtruppe* (*LFT*) also wanted the Friedrichshafen G.III, but their initial request in November 1917 was refused. In place of the Friedrichshafen G.III the *LFT* was offered the Gotha G.IV and purchased a batch of 40. However, Germany did release one Friedrichshafen G.III airframe to the *LFT* in April or May 1918 for the *LFT* to explore the possibility of license production. Meanwhile the Gotha G.IV(LVG) bombers in *LTF* service were experiencing severe problems that rendered them virtually useless, and the *LFT* again requested 50 Friedrichshafen G.III bombers. This time the request was approved, but a large fire at the Friedrichshafen factory on the night of April 13/14, 1918 precluded fulfilling the order. Due to increased license production by Daimler and Hansa, *Idflieg* finally approved the sale of 20 Friedrichshafen G.IIIa(Daim) bombers to the *LFT* on August 1, but none were delivered. On May 18, 1918 the Austro-Hungarian *Fliegerarsenal* awarded a contract for 100 bombers to the Oesterreichishe Flugzeugfabrik AG (Oeffag), 50 of which were to be the Friedrichshafen G.IIIa, which the *LFT* regarded as the most effective German two-engine bomber at that time. Two Friedrichshafen G.IIIa aircraft with Austro-Hungarian engines installed were sent as pattern aircraft. These aircraft, G.IIIa 789/17 and 790/17, with 250 hp Benz Bz.IV(Mar) engines, were shipped by train and arrived at the Oeffag factory in August. The German drawings had to be revised to Oeffag standards and final assembly of the prototype, Friedrichshafen G.IIIa(Oef) 54.01 began in September. However, with the Austro-Hungarian Empire crumbling, none of these bombers were completed before the Armistice.

Below: Friedrichshafen G.IIIa G.789/18 at Oeffag in August 1918; this aircraft had 250 hp Benz Bz.IV(Mar) engines built in Austria instead of the usual 260 hp Mercedes. The G.IIIa featured a number of improvements over the G.III. The most obvious was the 'box' tail to improve directional control with one engine inoperative. Other improvements included a 'Gotha tunnel' to improve the rear gunner's downward field of fire, greater fuel capacity for longer range, greater bombload, and Flettner servo-tabs on the ailerons to reduce the control forces and thus the pilot's workload.

Left: The passage allowing movement between the rear cockpit and the pilot's cockpit was introduced on the Friedrichshafen G.IIIb. The forward fuselage of this aircraft has not yet been completed.

Below: Friedrichshafen G.IIIa 1056/18 in Allied hands.

Below: Friedrichshafen G.III 179/17 was covered with the typical dark, night bomber printed camouflage fabric.

Friedrichshafen G.III

Friedrichshafen G.III 180/17 aircraft '4' of *Bogohl* I, *Bosta* 2, downed in Belgian lines in February 1918. It was later painted in Belgian markings, including the Belgian royal coat of arms on the nose. It wears typical night camouflage fabric with two-tone blue mottle on ply and metal surfaces. All crosses were outlined.

Friedrichshafen G.III 369/17 aircraft '5' of *Bogohl* 8, *Bosta* 25, crewed by *Lt.* Reinlein, *Lt.* Wetzlar, and *Vzfw.* Christl at Maria Alter Aerodrome, May 1918.

Friedrichshafen G.III

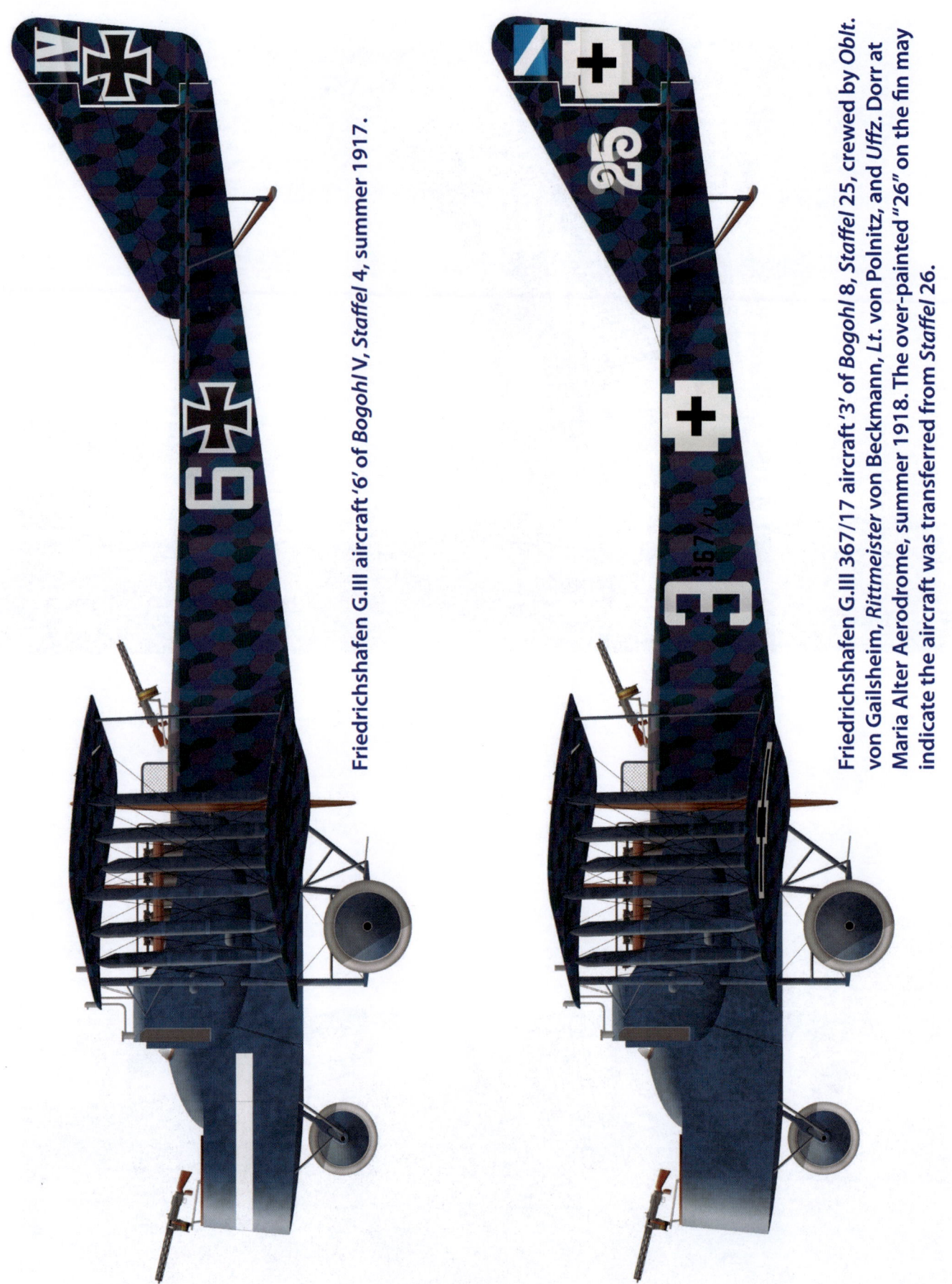

Friedrichshafen G.III aircraft '6' of *Bogohl* V, *Staffel* 4, summer 1917.

Friedrichshafen G.III 367/17 aircraft '3' of *Bogohl* 8, *Staffel* 25, crewed by *Oblt.* von Gailsheim, *Rittmeister* von Beckmann, *Lt.* von Polnitz, and *Uffz.* Dorr at Maria Alter Aerodrome, summer 1918. The over-painted "26" on the fin may indicate the aircraft was transferred from *Staffel* 26.

Above: Friedrichshafen G.III and G.IIIa bombers of *Bogohl* I in 1918.

Right & Below: Friedrichshafen G.III 180/17 of *Bogohl* I, *Bosta* 2 was brought down near Ghent, Belgium in February 1918.

Above: Friedrichshafen G.III 180/17 of *Bogohl* 1, *Bosta* 2 after capture in February 1918 with Belgian insignia applied.

Above: Friedrichshafen G.IIIa 1479/18 without engines, probably on display in Hyde Park, London.

Above: Friedrichshafen G.III being loaded with 50 kg bombs. The radiator details and forward gunner's position are nicely shown. A pintle mount for a second flexible forward-firing machine gun is mounted between the cockpits.

Right: Friedrichshafen G.III 180/17 under evaluation at the French test center at Villacoublay. The nosewheel helped prevent nose-overs during landing, one factor that made it safer to fly than Gothas. Another factor in flight safety was the center of gravity was in the right place, not aft like the Gotha bombers.

Above: Friedrichshafen G.IIIa 862/18 of *Bogohl* 8 in October 1918 during a flight for parachute testing.

Below: Friedrichshafen G.III with aircrew and ground-crew members during bomb loading demonstrating a 300 kg P.u.W. bomb.

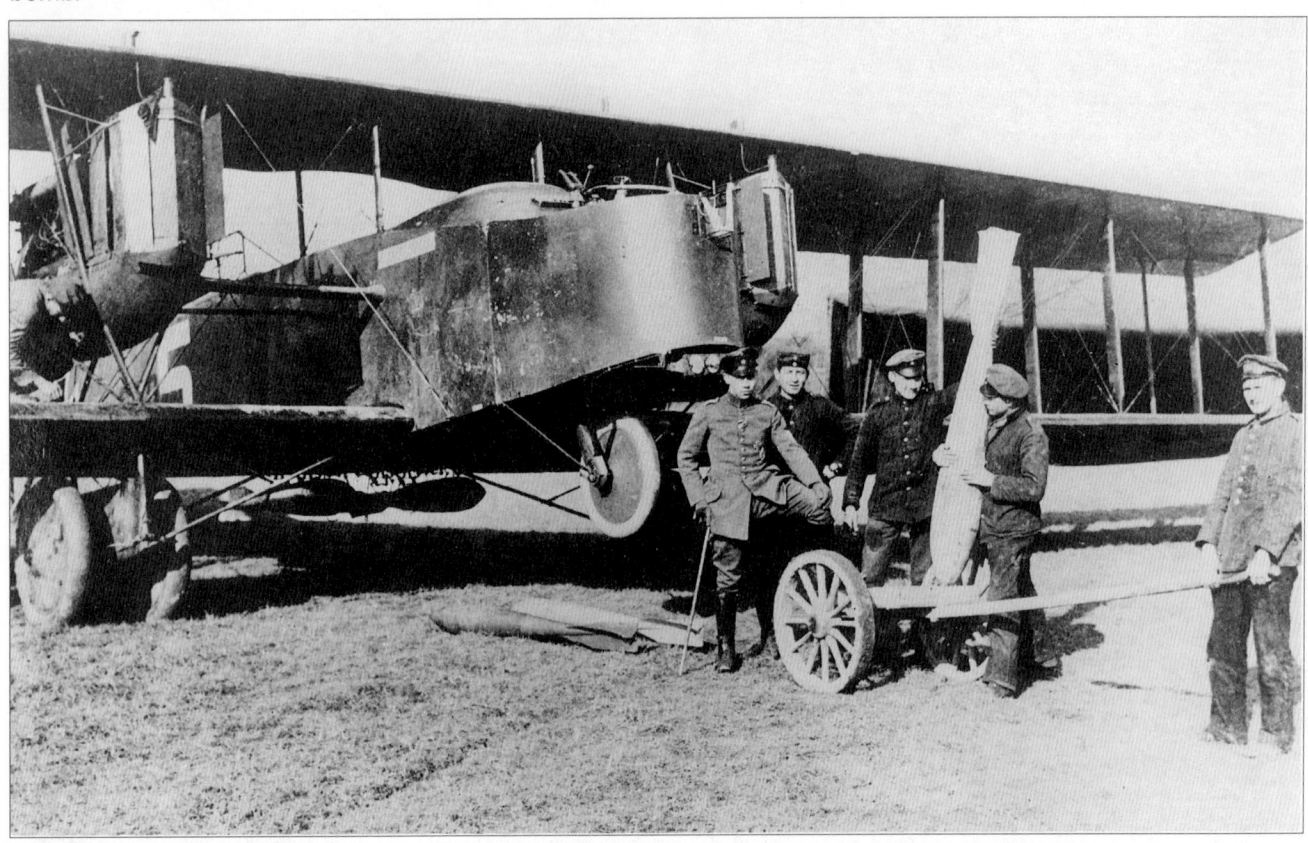

Friedrichshafen G.III & G.IIIa

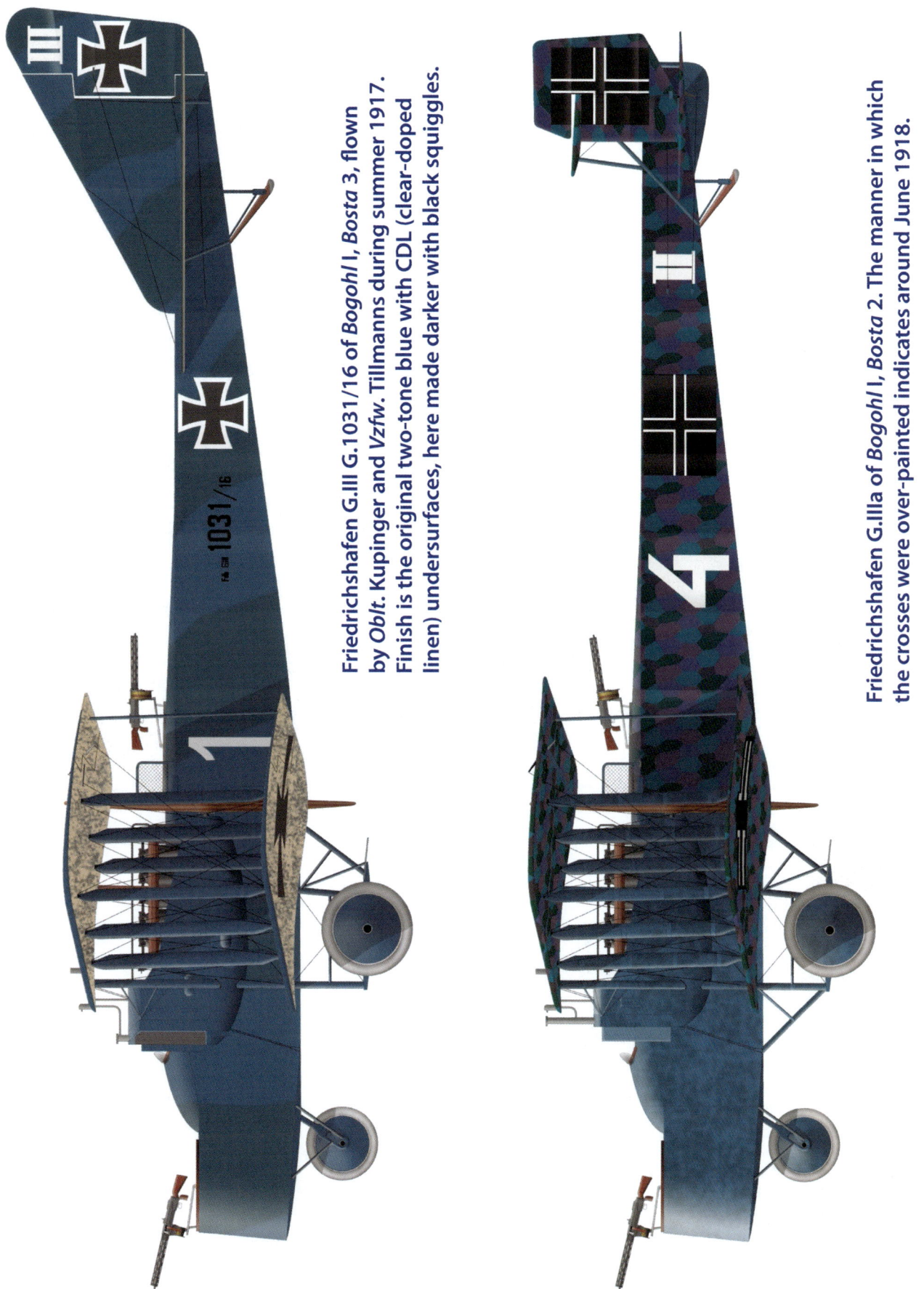

Friedrichshafen G.III G.1031/16 of *Bogohl* I, *Bosta* 3, flown by *Oblt.* Kupinger and *Vzfw.* Tillmanns during summer 1917. Finish is the original two-tone blue with CDL (clear-doped linen) undersurfaces, here made darker with black squiggles.

Friedrichshafen G.IIIa of *Bogohl* I, *Bosta* 2. The manner in which the crosses were over-painted indicates around June 1918.

Friedrichshafen G.IV

The Friedrichshafen G.IV, factory designation FF61, was a development of the Friedrichshafen G.IIIa to carry a larger bomb load. The G.IV retained the two 260 hp Mercedes D.IVa engines and box tail of the G.IIIa coupled with stronger four-bay wings. According to one reference 100 were ordered from Friedrichshafen and 35 from Daimler. The Caspar Hanseatische Flugzeugwerke was also to produce the type but apparently none were built before the Armistice. The G.IV was the penultimate development of the Friedrichshafen bomber family, the last to reach operational service, and the last to closely resemble the earlier bombers in the series, especially the G.IIIa.

Above & Below: The Friedrichshafen G.IV was the final Friedrichshafen bomber type to see service. It was a refinement of the G.IIIa designed to carry a heavier bomb load and had four-bay wing bracing.

Above: This is another photo of the Friedrichshafen G.IV on the facing page. The only significant difference between the G.IIIa and G.IV was the G.IV had a modified wing with four bays of bracing struts instead of the three bays of the G.IIIa.

Above: Another view of Friedrichshafen G.IV(Daim) 1074/18 also shown on the facing page. This aircraft was from the final batch ordered from Daimler in October 1918 and appears incomplete; perhaps construction was stopped due to the Armistice.

Above: Based on the use of tractor engines, this rear view appears to show the tail details of the Friedrichshafen G.V.

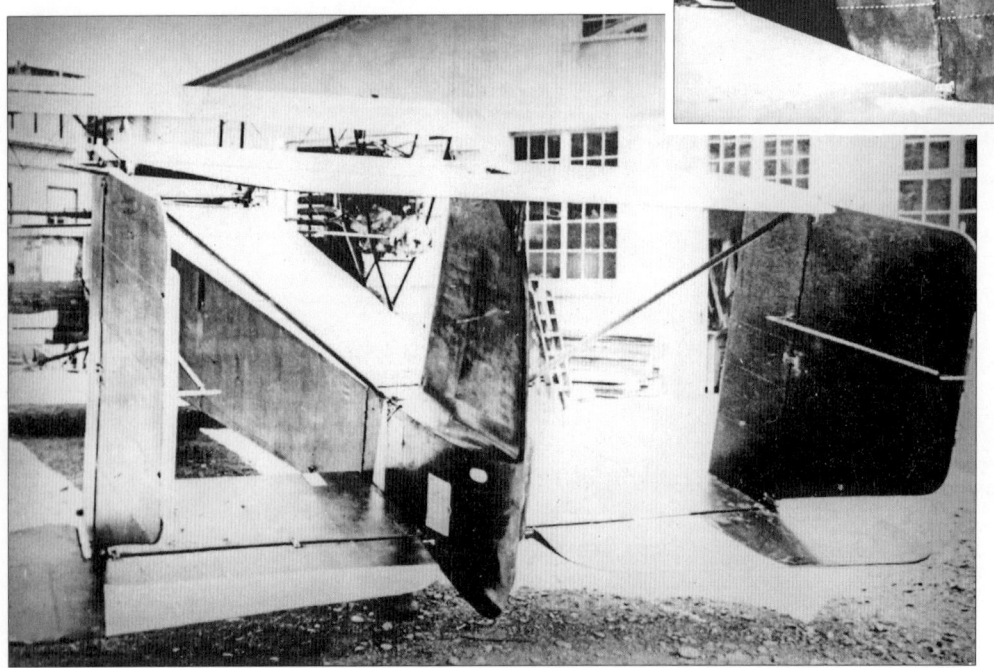

Above & Left: Experimental tail surfaces tested for the Friedrichshafen G.IIIa shows the fixed center fin that was tested but did not reach production. The rudders also appear to have longer chord than the production rudders. Designers constantly tried new tail designs to improve flying qualities and controllability after an engine failure.

Friedrichshafen G.IIIa

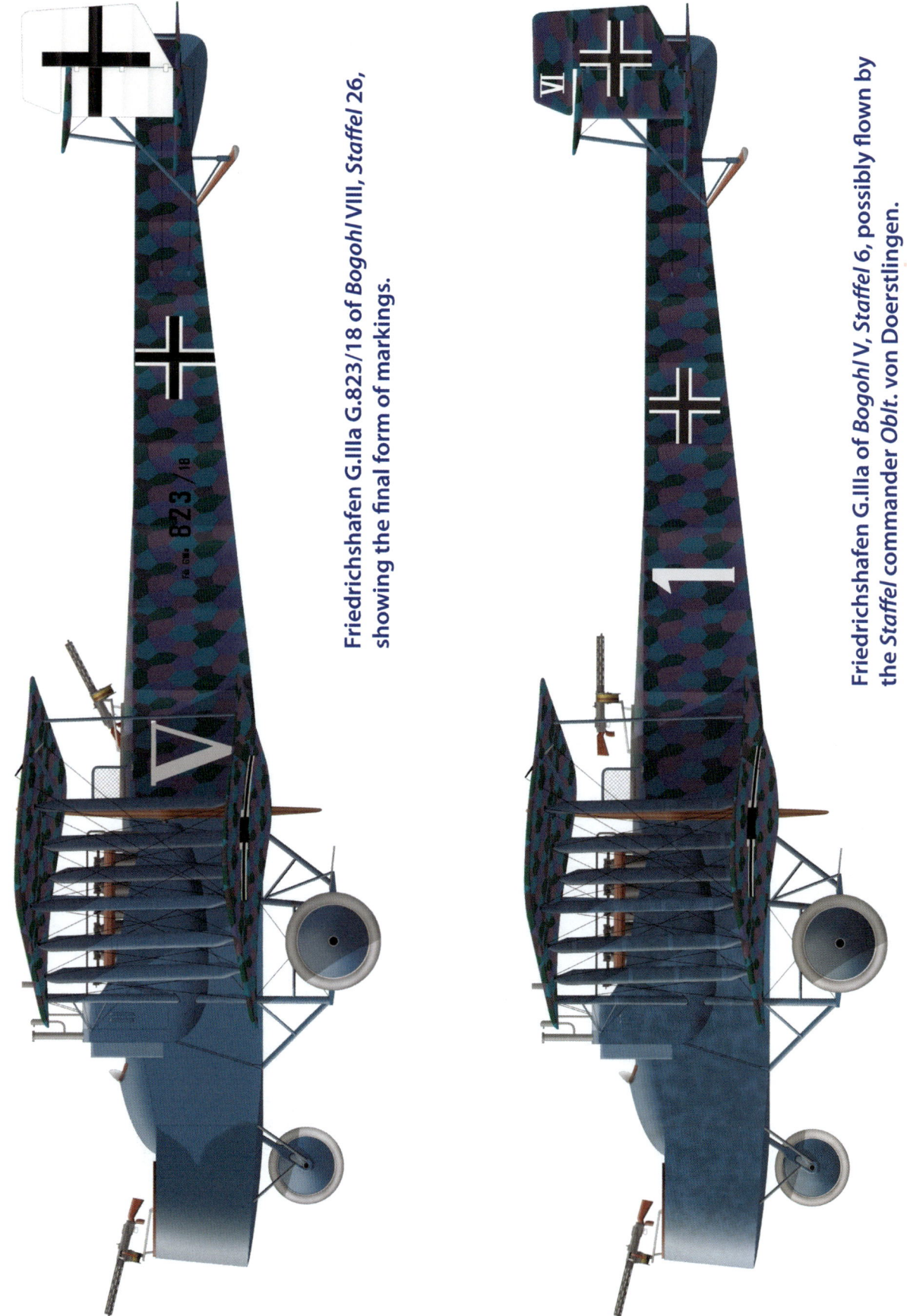

Friedrichshafen G.IIIa G.823/18 of Bogohl VIII, Staffel 26, showing the final form of markings.

Friedrichshafen G.IIIa of Bogohl V, Staffel 6, possibly flown by the Staffel commander Oblt. von Doerstlingen.

Friedrichshafen G.V

The Friedrichshafen G.V was the final development of the basic Friedrichshafen bomber family. Based on the G.IIIa airframe with box tail using a central fin, to further improve safety as a night bomber, the nose turret was removed and the engines were mounted in tractor configuration. Elimination of the nose gun position enabled the engines to be mounted closer together, reducing asymmetric thrust in case of engine failure and making the bomber easier to fly in that situation, which was important for flight safety during night operations. Two versions were built that differed in the type of engine used, factory designation FF55 being powered by two 245 hp Maybach Mb.IVa engines and factory designation FF62 using two 260 hp Mercedes D.IVa engines. Other than their engines, the two versions appear essentially identical.

According to one reference three examples of the FF55 were built and this version was the Friedrichshafen G.V. Why the reliable Mercedes used in the earlier G.III/IIIa and G.IV was to be replaced by the Maybach engine, which was more suited to high-altitude performance, for this low-altitude night bomber is unclear and seems counter-intuitive. However, first flight of the FF55 was on May 9, 1918, whereas first flight of the FF62 was in late November 1918, after the Armistice, and that may be the reason the Maybach-powered version was considered the G.V. Apparently three examples of the FF55 and only one FF62 were built before the Armistice, and the G.V was too late to see operational service.

Above: Photographed May 18, 1918, the Friedrichshafen G.V represented a significant modification of the basic Friedrichshafen bomber airframe. The nose turret was removed and the engines were mounted in tractor configuration, enabling them to be moved closer to the aircraft's centerline. This reduced asymmetric thrust after an engine failure, making the aircraft easier to control. Good handling qualities were vital to flight safety, especially during night landings, and the G.V also featured Flettner tabs on the ailerons to reduce the control forces, making flight less tiring for the pilot. The box tail featured a fixed central fin, which was tested but not used on earlier production Friedrichshafen bombers. The Mercedes engines were also replaced by 245 hp Maybach Mb.IVa engines, although the basic airframe was also tested with the 260 hp Mercedes D.IVa used in the earlier G.III and G.IV. Only three aircraft with Maybach engines (designated FF55) and one with Mercedes engines (designated FF62) were completed before the Armistice, so the G.V was too late for operational service. Photographs showing G.V 900 and G.V 901 are probably G.900/17 and G.901/17. The prominent drag-producing radiators indicate that speed was not a high prioirity for night bombers. The brace supporting the tail has been retouched out of the photo. (Peter M. Bowers Collection, Museum of Flight)

Above & Below: A Friedrichshafen G.V prototype carrying a 1,000 kg P.u.W. bomb photographed on May 9, 1918, the day of its first flight. Doubled wheels are fitted to support the additional weight. Unlike the Friedrichshafen G.IIIa and G.IV, the box tail now has a fixed fin. Like AEG and Gotha, Friedrichshafen continuously experimented the tail configuration of its bombers in a search for improved handling qualities, especially after an engine failure. The brace supporting the tail has been retouched out of the photo. (Peter M. Bowers Collection, Museum of Flight)

Notes on Friedrichshafen Bomber Serials by Reinhard Zankl

Confirmed Fdh G.IIIb aircraft:	G.1465/18, G.1466/18, G.1470/18, G.1471/18, G.1472/18 (the only confirmed serials in the G.1465–1564/18 range)
Confirmed Fdh G.IV aircraft:	G.508/18, G.526/18, G.827/18, others known in this series are G.IIIa or G.IV
Confirmed Fdh G.IV(Daim) aircraft:	G.1071/18, G.1091/18, G.1093/18, others known in this series are G.IIIa(Daim) or G.IV(Daim)

Gotha Bombers

Gotha is best known for its series of G-types, or twin-engine bombers. Although these were used initially as short-range tactical bombers, the Gotha G.IV had sufficient range to permit bombing Britain from forward air bases along the Belgian coast. It was these bombing raids, and especially the daylight raids on London that started on 13 June 1917, that made Gotha a household name, and one to be feared.

Although damage from these raids was relatively minor, public outrage forced the British government to substantially build up the defenses of London; fighter squadrons were relocated to London from the Western Front and the anti-aircraft gun batteries around London were strengthened. The heavy guns assigned to London could not be used on the Western Front nor installed on ships to protect against surfaced U-boats.

In a matter of a few months the daylight anti-aircraft defenses of London were strong enough that the Gothas were forced to turn to night bombing operations over Britain, where they were soon joined by the newer Gotha G.V as well as Giant Staaken bombers. The night attacks were also of limited military value, but the need for Britain to mount a strong defense wasted resources that would have been better employed on the Western Front.

Bombs were normally loaded well forward of the Gotha bombers' center of gravity, but once the bombs were dropped the Gothas became tail heavy and pitch stability greatly deteriorated, making them difficult to fly and causing crashes on landing, normally the most dangerous part of flying. Landing crashes caused 76% of all Gotha bomber losses, more than three times losses from all other causes combined.

After the last bombing raid on Britain the night of 19/20 May 1918, the Gotha bombers returned to short-range tactical night bombing.

Above: Gotha G.IV of *Kagohl 3*, the *England Geschwader*, the unit assigned to bomb Britain. There are bombs under the extreme nose of the airplane; this unusual placement was due to the tail-heaviness of the Gotha bombers. Once the bombs were released the Gothas became neutrally stable or unstable in pitch, resulting in numerous landing accidents that destroyed more Gotha bombers than all other causes combined. At this remove it is difficult to understand why Gotha did not fix the tail-heaviness problem before the bombers were produced in quantity. Regardless, these bombers were Gotha's major success as an aircraft manufacturer. Without them Gotha would have little reputation in aviation.

Gotha G.I

At the dawn of aviation, the military authorities of the major powers considered how aviation might be used in war. Reconnaissance and bombing immediately suggested themselves as potential roles for aviation, and in fact both were important in WWI and since. Of course, the next thought was, how to defend against the enemy using his aviation assets against your own forces? This concern led to anti-aircraft guns and the idea of using airplanes for air-to-air combat.

But what kind of airplane would be best for defeating other aircraft? In retrospect the answer is clear, but it was far from obvious before the war. Synchronizers to enable machine guns to fire between the blades of a rotating propeller were being designed, but got little notice at the time. The concept that resonated with many designers and military authorities before the war, before there was any experience in air-to-air combat, was the idea of aerial cruisers. The aerial cruiser theory was derived from the warships of the day; the pilot would fly the aerial cruiser within range of the enemy aircraft and gunners would fire flexibly-mounted machine guns and cannon to destroy the adversary. Thus was born the idea of the battleplane, and Britain, France, and Germany all built aircraft to this concept.

Friedel-Ursinus B.1092/14 Specifications

Engines:	2 x 100 hp Mercedes D.I	
Wing:	Span Upper	22.00 m
	Span Lower	19.00 m
	Chord (Upper & Lower)	2.2 m
	Sweepback	4°
General:	Length	17.60 m
	Height	6.00 m
	Empty Weight	4700 kg
	Loaded Weight	6860 kg
Maximum Speed:		90–95 km/h

Gotha G.I Specifications

Engines:	2 x 150 hp Benz Bz.III	
Wing:	Span Upper	20.30 m
	Span Lower	19.70 m
	Chord (Upper & Lower)	2.20 m
	Gap	1.95 m
	Sweepback	10°
	Area	82 m²
General:	Length	12 m
	Height	3.9 m
	Empty Weight	1800 kg
	Loaded Weight	2966 kg
Maximum Speed:		130 km/h
Climb:	2000m	47 min

Above: The modified Friedel-Ursinus B.1092/14 and a captured Morane monoplane at *FEA* 9 in Darmstadt for a size comparison. Aerodynamically-balanced ailerons have been fitted and the nose radiators have been replaced by larger, twin radiators on each engine. A long cellon window is in the fuselage side over the wing.

Above: Three Gotha G.I aircraft stand on the Gotha airfield ready for flight testing. These may have been the first three built, G.9/15, G.10/15, and G.11/15, all of which were sent to *FEA* 7 to defend the Krupp steel works in Cologne.

Below: Closeup of the first Gotha G.I in the lineup above. The national insignia were painted below the top wing in addition to the usual locations. Both photos were released as Sanke cards.

Above: Gotha G.I 13/15 was delivered to *FEA* 3 on 2 September 1915, then flown to the Eastern Front. Enroute it landed at Schneidemühl where this photograph was taken. A streamlined bomb container is just visible between the wheels.

In early 1914 key German aviation authorities, including *Idflieg*, the *VPK* (*Verkehrstechnische Prüfungs Kommission* = transport technical investigation commission) and aviation industry executives, discussed the role of military aircraft and reached a consensus. The VPK then issued a directive outlining the tactical role of aircraft and specified three categories: *Typ I* was a fast two-seater intended for extended flights for reconnaissance and light bombing; *Typ II* was a light, maneuverable two-seater for short flights over the lines and armed for self-defense; and *Typ III* was three-seater designed to carry a large payload and fly low within range of enemy fire. The *Typ III* was required to have a speed over 120 km/h, climb to 800m in 10 minutes, have a flight duration of 6 hours, and a useful load of 450 kg. In essence the *Typ III* was a battleplane (*Kampfflugzeug* in German).

The German army approved the *VPK* recommendations on 28 April 1914, and demanded that these airplanes be developed as soon as possible. The *Typ I* was essentially the B-type two-seat biplane, of which numerous designs were available, while the *Typ II* eventually converged with the developed B-type to become the C-type armed two-seater. In parallel, a number of companies responded with designs to the *Typ III* requirements.

However, the design that was to evolve into the Gotha G.I stemmed from an independent effort.

Oskar Ursinus, founder and editor of *Flugsport* magazine and a civil engineer, received orders to report to *FEA* 3 (*FEA* for *Flieger Ersatz Abteilung* = aviation replacement unit) in Darmstadt on 1 August 1914. On 9 August Ursinus proposed building a twin-engine *Kampfflugzeug* to *Major* Friedel, *FEA* 3's new commander, using under-utilized military personnel from the unit. Friedel accepted and the new airplane would be known as the Friedel-Ursinus *Kampfflugzeug*, also known as the type FU. The aircraft was certainly built with the *Typ III* requirements in mind; for example the fuselage and engine nacelles were armored.

Design work started immediately, and on 1 September *FEA* 3 personnel began to build the aircraft, which was assigned the military designation B.1092/14. This indicates *Idflieg* approved the aircraft and may have provided funds. On 30 January 1915 pilot Herold performed the aircraft's first flight.

The aircraft was powered by two 100-hp Mercedes D.I engines. The high fuselage enabled the engines to be very close to the centerline, reducing asymmetric control forces in event of engine failure. It also gave a wide field of fire to the gunner in the nose. However, with no protective structure above the crew, a turnover on landing would be extremely dangerous to the crew.

According to Ursinus's biography, the type FU was eventually sent to Ujatz, near Lodz, for operational

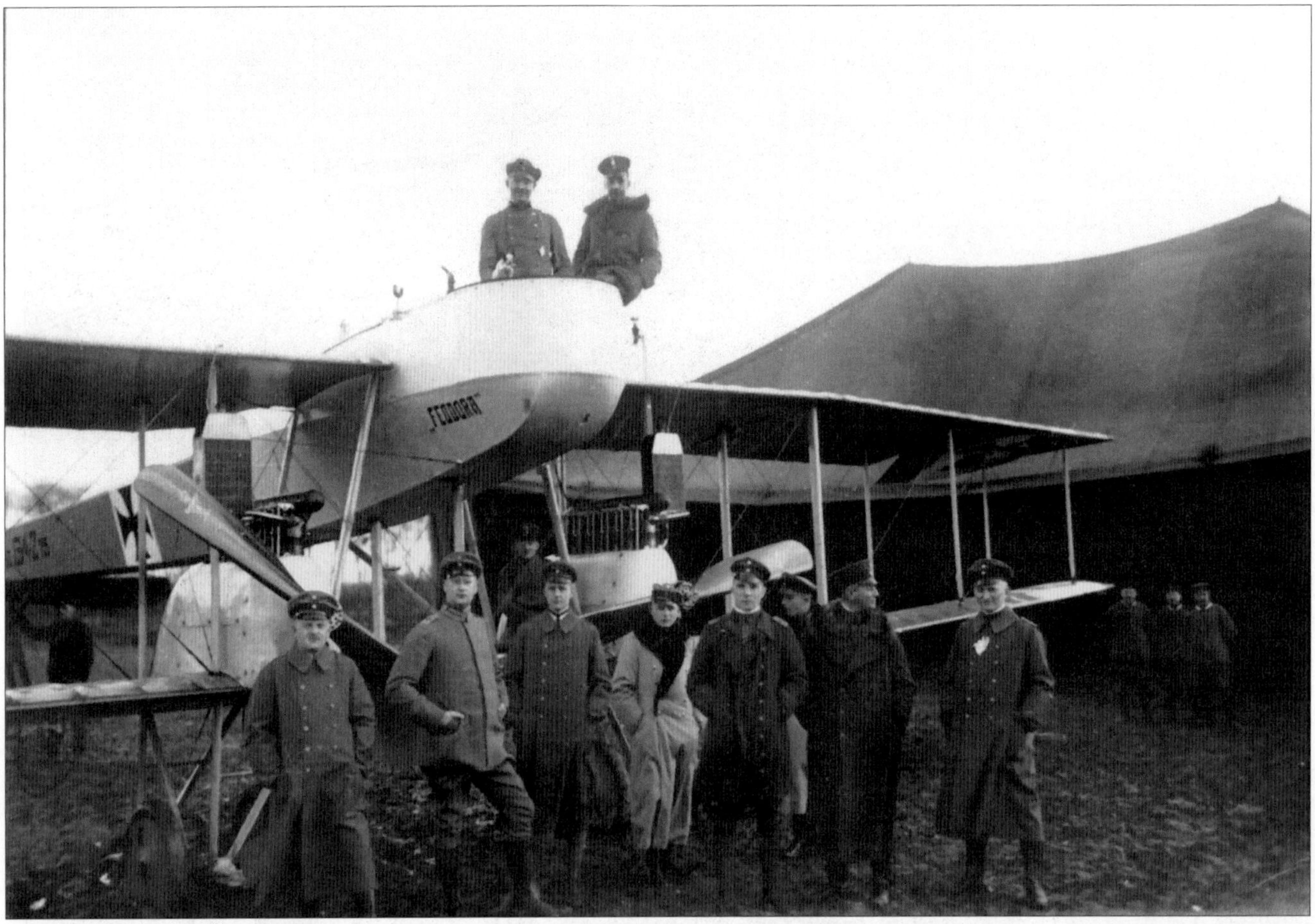

Above: Gotha G.I 42/15, named *Feodora*, was delivered to *FEA* 3 in Autumn 1915 before being sent to the Eastern Front.
Below: Gotha G.I 43/15 of the second production series.

Gotha G.I

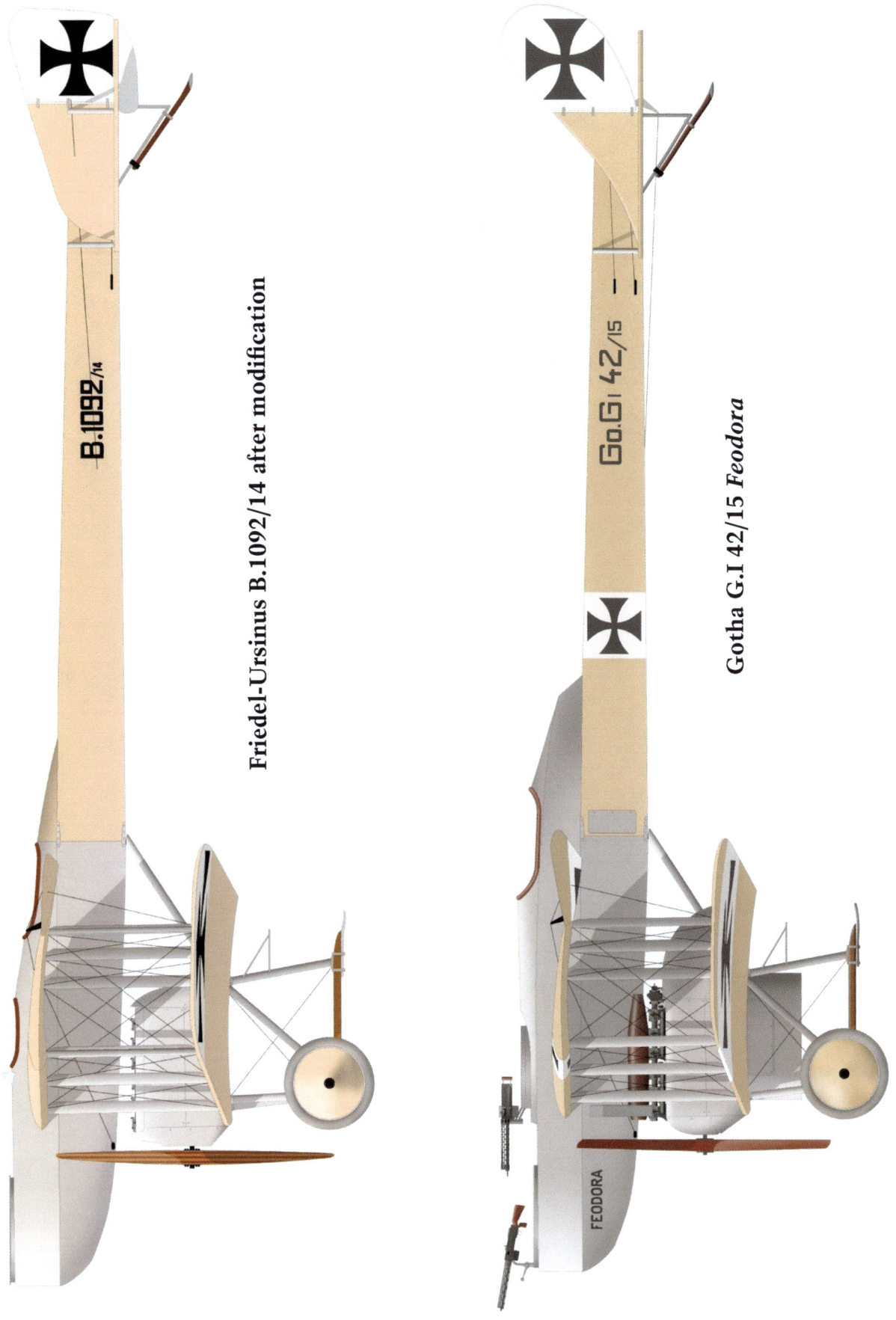

Friedel-Ursinus B.1092/14 after modification

Gotha G.I 42/15 *Feodora*

Left: Oskar Ursinus sits on a ladder in front of a Gotha G.I. Ursinus, founder and editor of *Flugsport* magazine and a civil engineer, designed the original known as the Friedel-Ursinus *Kampfflugzeug*, and Gotha produced a modified version of that design as the Gotha G.I.

Right: Oskar Ursinus in the front gunner's cockpit of a Gotha G.I. The variety of photos of Ursinus handling the front gun in a G.I seems to indicate an aggressive nature, but Ursinus is not known to have flown in combat.

Left: This may be Gotha G.I G.41/15 of the second production batch, but the serial number is partly obscured.

Above: Engines running, a Gotha G.I is ready for take-off.

trials on the Russian Front, but there is no further information.

A production license for the type was offered to Fokker and Gotha; Gotha signed a license in March 1915. Likely in anticipation of that event, in February 1915 *FEA 3* was transferred to Gotha. On 1 April 1915 *Idflieg* awarded a contract for six Gotha G.I aircraft to Gotha. The internal Gotha company designation was type UKL or type GUK; both were used. Of the first batch of six aircraft, five were to be powered by the 150 hp Benz Bz.III and the other was to be powered by two 160 hp Mercedes D.III engines. The contract required a crew of two with one machine gun, 200 kg of bombs, 150 kg of armor, and a maximum speed of 125 km/h.

Ursinus worked with Gotha engineer Hans Burkhardt to prepare manufacturing drawings. Starting on 27 July the aircraft were delivered as G.9/15–G.14/15; the last was delivered on 8 September. The first three went to *FEA* 7 for defense of the Krupp works in Cologne, the next two went to *FEA* 3 at Gotha for training, and the last went to *Armeeabteilung* Falkenhausen. The production G.I differed in detail from the Type FU.

On 15 July 1915 a second series of six Gotha G.I aircraft were ordered, all to be powered by the 150 hp Benz Bz.III. These aircraft were accepted between 22 September and 5 November, and all remained at *FEA* 3 in Gotha.

The third and final series of six Gotha G.I aircraft was ordered on 17 October 1915. These were powered by the 160 hp Mercedes D.III engine. The *Kampfflugzeug* concept now having failed, bombing was emphasized and the required bomb load was raised to 350 kg. A third crewman with a second gun carried between the pilot and front gunner was now specified. *Idflieg* also asked that a machine cannon be installed in addition to a machine gun, leading to weapons trials with a 20 mm Becker cannon and 37 mm cannon. The final batch was delivered between 24 January 1916 and 20 March 1916. One machine went to *FEA* 1 at Döberitz; the other five went to the *Prüfanstalt und Werft* (*Idflieg's* test establishment and workshop) at Döberitz.

In service the slow G.I accomplished little as a battleplane, the basic concept being fundamentally flawed. Battleplanes of all designs soon demonstrated they were too slow to catch enemy aircraft.

Above: The battleplane heritage of the Gotha G.I is demonstrated here. Oskar Ursinus handles a 20mm Becker cannon in the nose turret while the second gunner demonstrates the Parabellum machine gun; the pilot is seated aft.

Furthermore, even when the enemy was within range, so were they; the enemy had as good a chance at victory as they did. Operational experience soon showed that large, twin-engine types were better suited for bombing than air superiority missions, leading to the last batch of G.Is being modified for a greater bomb load. More specifically to the unique design the G.I inherited from the Type FU, turnovers on landing were always a possibility and, if they occurred, were invariably fatal to the crew. Moreover, the G.I was structurally fragile.

The German Navy purchased one float-equipped version of the G.I as the Gotha UWD, Marine Number 120.

Gotha G.I Production Summary				
Production Batch	**Serial Numbers**	**Qty**	**Engines**	**Crew/Armament**
#1 (ordered 1-4-15)	G.9/15–G.14/15	6	2x150 hp Benz Bz.III (five) 2x160 hp Mercedes D.III (one)	2 crew, 1 gun, 250 kg bombs
#2 (ordered 15-7-15)	G.40/15–G.45/15	6	2x150 hp Benz Bz.III	2 crew, 1 gun, 250 kg bombs
#3 (ordered 17-10-15)	G.100/15–G.105/15	6	2x160 hp Mercedes D.III	3 crew, 2 guns, 350 kg bombs
The Friedel-Ursinus prototype, assigned designation B.1092/14, had one gun.				

Facing Page: Oskar Ursinus demonstrates the use of the forward gun in a Gotha G.I. The streamlined bomb container between the wheels is clearl y visible and the nose of a 20-kg Carbonit bomb just protrudes from the lower front. Together with the bomb-dropping chute beneath the gunner's cockpit the bomb container is a clear indication of the evolution of the battleplane into a bomber. The propellers rotate in opposite directions to minimize torque.

Gotha G.II

Construction of the Gotha G.I gave Gotha the experience needed to design and build a long-range bomber, something *Major* Wilhelm Siegert, the commander of *Idflieg* and a supporter of strategic bombing, had wanted since the beginning of the war. Gotha engineer Hans Burkhardt, who had worked with Oskar Ursinus to build the Gotha G.I, was the natural choice for designer of the new bomber.

Earlier Burkhardt had modified crashed Gotha G.I G.9/15 by placing the fuselage on the lower wing, which greatly reduced the likelihood of a nose-over on landing. Burkhardt stated he had three main priorities when he designed the Gotha G.II; speed, protection of the observer in the nose, and ease of transportation. The later sounds odd today, but at that time, when airplanes were transported long distances, they were normally dismantled and moved by train rather than flying them. This counter-intuitive procedure was driven by the limited reliability of contemporary aircraft.

Burkhardt's ideas were accepted by *Idflieg*, who placed a production order for ten bombers on 18

Gotha G.II Specifications		
Engines:	2 x 220 hp Mercedes D.IV	
Wing:	Span Upper	23.70 m
	Span Lower	21.90 m
	Area	89.5 m²
	Chord Upper	2.30 m
	Chord Lower	2.30 m
	Gap	2.22 m
	Sweepback	1.5°
General:	Length	12.40 m
	Height	4.30 m
	Empty Weight	2182 kg
	Loaded Weight	3192 kg
Maximum Speed:		148 km/h
Climb:	3000m	28 min
	4000m	41 min
Range:		500 km

December 1915. To have a worthwhile range and payload the new bomber needed much more power than the Gotha G.I, and the engines were specified as 220 hp Mercedes D.IV straight-eights mounted as pushers. The Mercedes D.IV was an inline eight-

Above: The Gotha G.II prototype was a completely different aircraft than the G.I. Distinguishing characteristics include the two-bay wing, four-wheel undercarriage module under each engine, and WWI too-small rudder with no fin.

Above: This side view of the Gotha G.II prototype gives a better view of the inadequate vertical tail surfaces. The control cables were let outside the rear fuselage, adding drag to the airframe.

cylinder motor developed from the reliable 160 hp Mercedes D.III six-cylinder by adding two more cylinders. The Mercedes D.IV was powerful and reliable in single-engine airplanes, but sometimes suffered crankshaft failures in twin-engine airplanes due to flexing of the block. A transitional design, the engine was soon replaced in production by the simpler six-cylinder D.IVa.

Above: At least five production Gotha G.II bombers are parked in front of the flight test hangar in which the VGO Giant aircraft were built. Nearest the camera is G.203/16 with two-blade propellers, then G.205/16 with four-blade propellers, with G.200/16 behind it. The greatly enlarged vertical tail with fixed fin and enlarged, three-bay wings of the production aircraft is clearly shown. The engine was the straight-eight 220 hp Mercedes D.IV.

Above: Gotha G.II 204/16 photographed at the Gotha factory. This was one of the aircraft delivered to *Kagohl* 4, *Staffel* 20 on 24 August 1916. The G.II airframe was the basis for the later G.III, G.IV, and G.V.

The resulting Gotha G.II was a completely new design that established the configuration for all subsequent operational Gotha bombers. The prototype G.II was a two-bay aircraft that entered flight trials in March 1916. The under-carriage had eight wheels, four underneath each nacelle, for safe landings. Both two-bladed and four-bladed propellers were tried with the slow-turning geared engine. Matching a fixed-pitch propeller with engine and airframe to maximize speed and climb was a painstaking task that could only be accomplished by time-consuming flight-testing.

The prototype G.II was fast for its time but had insufficient climb with a full bomb load. An enlarged, three-bay wing gave the production version the required performance.

The undercarriage was also revised for the production G.II. Although the original undercarriage prevented nose-overs, it lacked brakes as did most WWI aircraft. An extended landing run could result in a fatal accident by running off the edge of the field. To solve this potential problem Burkhardt simplified the undercarriage by removing the forward wheels and moved the center of gravity aft, allowing a conventional tail skid that provided the necessary braking action. Unfortunately, this aft shift in the center of gravity made the aircraft far less stable in pitch, a situation that was to plague all subsequent Gotha bombers in service. Landing accidents due to pitch instability were responsible for 76% of all Gotha bomber losses, more than three times as many losses as all other reasons combined.

Finally, the too-small rudder of the G.II prototype was replaced with a fixed fin and larger rudder in the production G.II, improving stability and controllability in flight.

The G.II had a crew of three; a bombardier-gunner in the front cockpit, the pilot in the center cockpit, and a gunner in the aft cockpit. The wings and fuselage were of typical wood and fabric construction, although the nose was covered with plywood. The tail surfaces were of steel tubing covered with fabric. The engine nacelles and undercarriage formed a single sub-assembly; for easy handling these could be moved around when fitted with separate, auxiliary wheels, a feature that

Above: Gotha G.II 207/16; the dark nose is due to being covered in plywood; the rest of the airframe is covered in fabric.

was patented. Fuel and oil tanks were in the engine nacelles with a gravity fuel tank mounted above the upper wing. Two bomb racks in the fuselage held fourteen 10 kg bombs. The total useful load was 1,010 kg.

Production of the G.II began on 25 April 1916 and the type test was completed on 17 July 1916. Serial numbers assigned to the ten production aircraft were G.II 200/16–209/16. The G.II wing cellule finally passed the static load test during the week of 11 August 1916 after six failures that had to be rectified.

Of the 10 G.II bombers built only eight reached *Staffel* 20 of *KG*4 in August/September 1916, apparently the only unit to use the G.II operationally. One G.II remained at *FEA* 3 in Gotha and the other was badly damaged during flight evaluation. Unfortunately, there is little information available regarding the combat use of the G.II, which was limited by the small number of aircraft and the problematical reliability of the engines.

Gotha G.III

The G.II was quickly superseded in production by the improved G.III. The main difference between the types was the engine; the G.III was powered by the 260 hp Mercedes D.IVa engine of six cylinders. The new engine did not suffer the occasional crankshaft failures of the straight-eight it replaced, was cheaper to produce, offered slightly more power, and was soon available in quantity. The additional power enabled the useful load to be increased from 1,010 kg to 1,235 kg.

The other important change for the G.III was an opening for a ventral, downward-firing machine gun. *Idflieg* ordered 25 G.III bombers on 3 May 1916; deliveries started on 16 October and were completed on 26 March 1917. Serial numbers are thought to be G.375/16–G.399/16. One G.III, G.398/16, was sent to *Halbgeschwader* I as a trainer and G.392/16 was sent to the Daimler factory for flight-testing an experimental Mercedes D.IVa. The rest of the G.IIIs were delivered to *KG*2. *KG*2 was moved around and flew missions on the Western Front as well as daylight bombing missions on the Balkan Front, stationed at Hudova along with a few G.II bombers. The most significant result in the Balkans was destruction of the railway bridge over the Donau at Cernovoda in late September 1916; this deprived the Romanian army of essential supplies and reinforcements. Interestingly, one squadron commander complained that the G.III easily outdistanced its two-seat escorts. In August–September 1917 the G.III was retired from the front.

Gotha G.III Specifications		
Engines:	2 x 260 hp Mercedes D.IVa	
Wing:	Span Upper	23.70 m
	Span Lower	21.90 m
	Area	89.5 m²
	Chord Upper	2.30 m
	Chord Lower	2.30 m
	Gap	2.22 m
	Sweepback	1.16°
General:	Length	12.22 m
	Height	3.90 m
	Empty Weight	2383 kg
	Loaded Weight	3618 kg
Maximum Speed:		140 km/h
Range:		700 km

Above: Gotha G.III 376/16 or 378/16; like the G.II, the dark nose is due to being covered in plywood; the rest of the airframe is covered in fabric.

Below: Gotha G.III; the bulge on the upper side of the cockpit provided room for the pilot's controls.

Facing Page: Gotha G.III 398/16; differences between the G.II and G.III were slight and only visible from certain viewpoints.

Gotha G.III

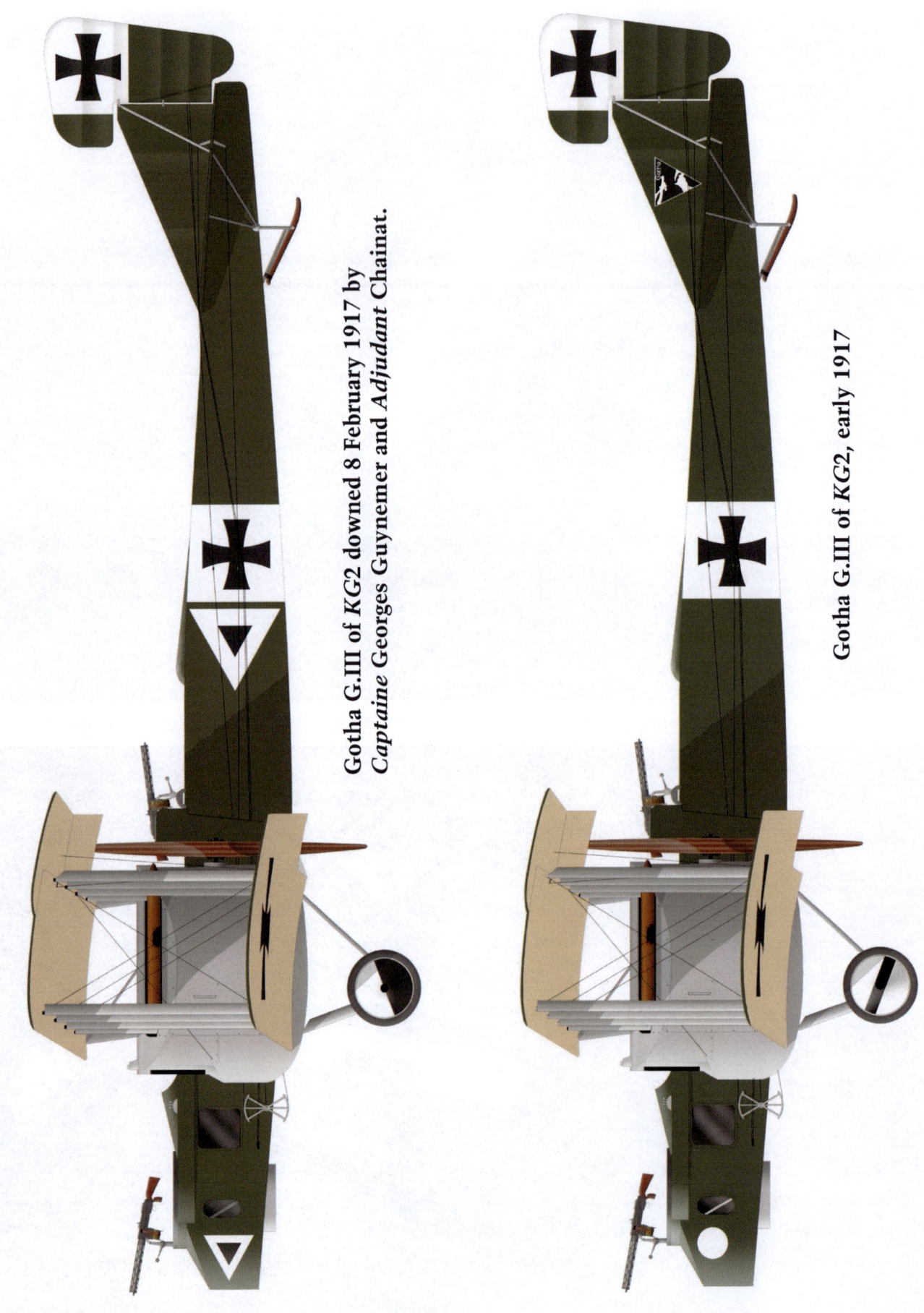

Gotha G.III of *KG2* downed 8 February 1917 by Captaine Georges Guynemer and *Adjudant* Chainat.

Gotha G.III of *KG2*, early 1917

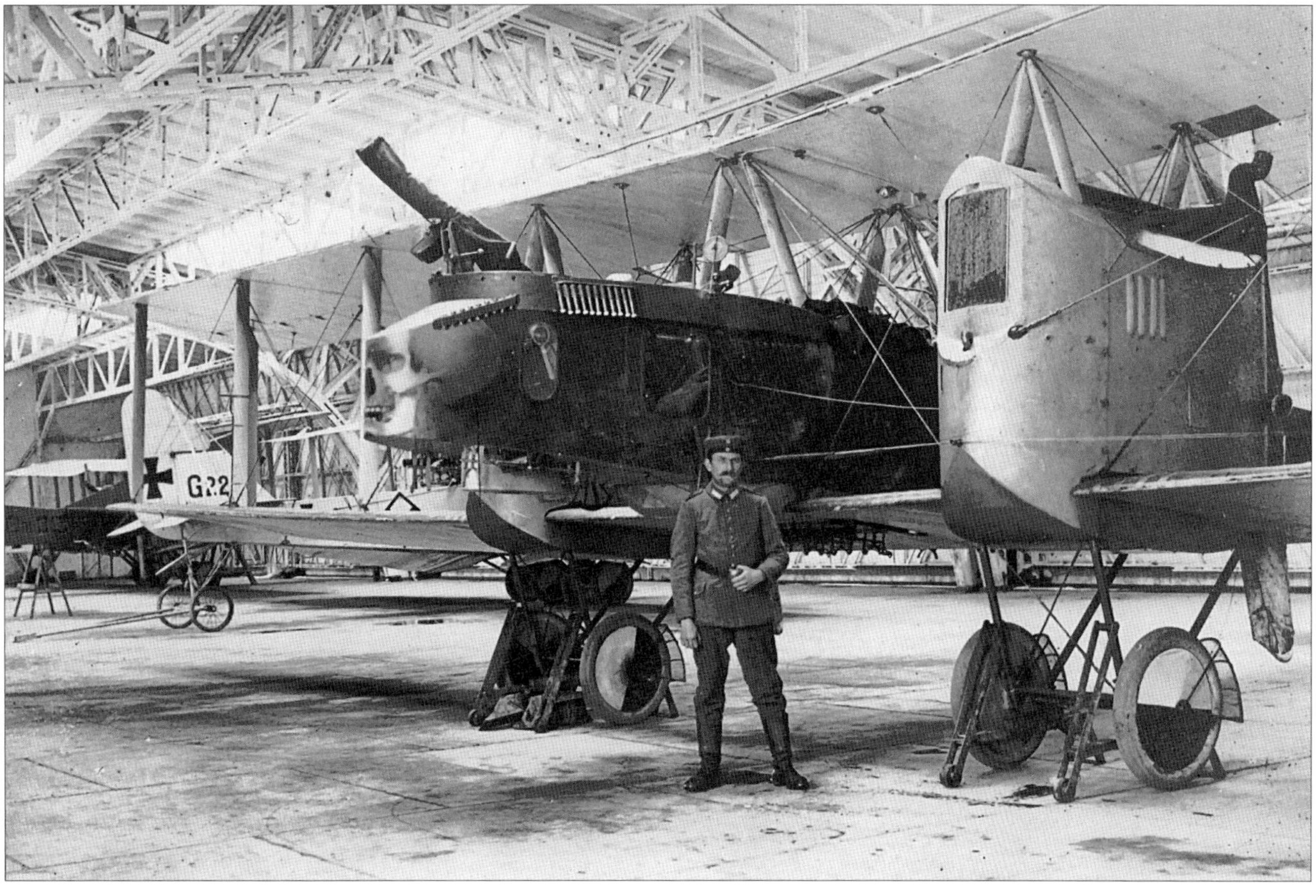

Above & Below: This Gotha G.III 389/16 (or possibly 385/16) of *Kagohl* 2, *Staffel* 19 wears a macabre 'death's head' marking on the nose. The upper surface and sides of the fuselage and upper surfaces of the wings were painted green. The outer wheel covers were painted in the black and white halves typical of *Kagohl* 2. A black triangle with white outline is painted on the rear fuselage. Belts of signal flares are attached to the side of the bombardier's cockpit, and the bomb racks are visible under the wing center section.

Above & Below: Gotha, probably a G.III but possibly a G.II, on a peaceful flight displays the characteristic Gotha shape.

Gotha G.III

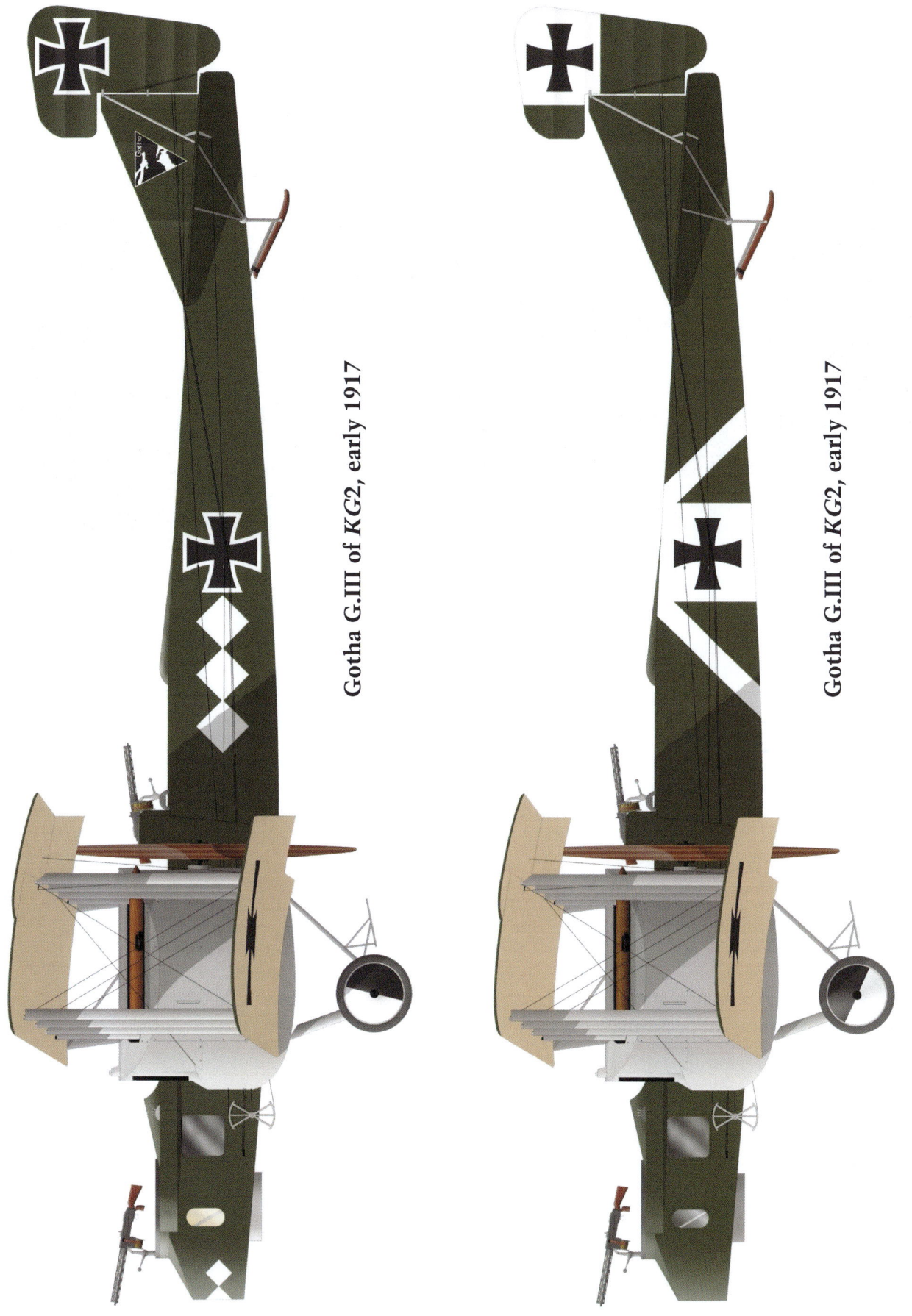

Gotha G.III of *KG2*, early 1917

Gotha G.III of *KG2*, early 1917

Gotha G.IV

By late 1916 it was apparent to the German Army that the strategic bombing campaign against Britain using Zeppelins was not a success, and furthermore the Zeppelins were too costly and vulnerable. The Army was planning to abandon airships in favor of cheaper, more effective heavy bombers. The operation to attack London and other strategic targets with bombers was called *Türkenkreuz* (Turk's Cross).

Of course, the plan required bombers with sufficient range and payload for the task, and the Gotha's performance made it the obvious choice. On 16 August 1916 *Idflieg* ordered 52 Gotha-built bombers of an improved type, the G.IV, making the G.IV the first Gotha design to be ordered in substantial numbers. So many of the new Gotha G.IV bombers were wanted that two other manufacturers, LVG and SSW, were given contracts to build the G.IV under license.

The G.IV was simply a refinement of the G.III. Powered by the same 260 hp Mercedes D.IVa engine used in the G.III, the G.IV featured ailerons on both upper and lower wings connected by an actuating strut for improved controllability. The G.IV also introduced the patented Gotha tunnel, a hollowed-out opening in the rear fuselage that enabled the gunner to depress his gun through the tunnel to fire at aircraft below, or to mount a ventral gun at floor level for a wider field of fire. To retain torsional rigidity despite the cut-out in the bottom of the fuselage, the fuselage was covered in plywood instead of fabric.

Gotha G.IV Specifications		
Engines:		2 x 260 hp Mercedes D.IVa
Wing:	Span Upper	23.70 m
	Span Lower	21.90 m
	Area	89.5 m²
	Chord Upper	2.30 m
	Chord Lower	2.30 m
	Gap	2.22 m
	Sweepback	1.16°
General:	Length	12.40 m
	Height	4.30 m
	Empty Weight	2413 kg
	Loaded Weight	3648 kg
Maximum Speed:		140 km/h
Climb:	1000m	3 min
	2000m	9 min
	3000m	16.5 min
	4000m	25 min
Range:		700 km

Gotha G.IV(LVG)

LVG received an order for 150 Gotha G.IV(LVG) bombers in December 1916. The type test for the initial 50 aircraft was completed on 25 June 1917, after the Gotha-built aircraft had already bombed London. One aircraft was fitted with two 245 hp Maybach Mb.IVa engines in an attempt to raise the bombers' operational altitude, but inability to find a suitable propeller thwarted the attempt. Another

Below: Gotha G.IV 408/16 was a bomber assigned to *Kagohl* 3, the *England Geschwader*, the unit assigned to bomb Britain. Two fuel tanks are mounted above the upper wing to give it enough range for these missions.

Above: Gotha G.IV 408/16 of *Kagohl* 3 after full tactical markings were applied. The letters on the fuselage side are likely the initials of two of the crew members, a common marking practice in *Kagohl* 3.

aircraft tested Flettner servo tabs starting in October 1917.

LVG built an additional 40 G.IV bombers to an Austro-Hungarian order. These were modified to take 230 hp Hiero engines built in Austria. The squadrons received these aircraft in March–April 1918. Weak engine bearers and unsuitable propellers led to excessive engine vibration that caused piping leaks and structural damage, limiting their usefulness, and by September 1918 the G.IV(LVG) bombers were essentially grounded.

Gotha G.IV(SSW)

Idflieg ordered 80 Gotha G.IV(SSW) bombers from SSW at the same time the LVG machines were ordered. The first batch of 40 were delivered between July 1917 and February 1918; about 30 went to operational units and the rest went to training and replacement units. The second production batch, ordered in May 1917, was delivered to training units between December 1917 and August 1918. The twin-wheel *Stossfahrgestell* (shock landing gear) designed by Siemens was fitted to all aircraft from G.217/17, and many machines were also fitted with the Flettner servo tabs. Instead of the normal 260 hp Mercedes D.IVa engines, some aircraft had 185 hp NAG C.III or 180 hp Argus As.III engines for training use.

Gotha G.IV Operations

The G.IV was the aircraft that made Gotha a household name. Built in much greater numbers than earlier models – 232 were built by Gotha, LVG, and SSW – it was intended by the German Army to replace the expensive, vulnerable Zeppelins as Germany's long-range bomber of choice. Although the first G.IVs were delivered in November 1916, a number of problems had to be resolved before the G.IVs were ready to bomb London, their intended target, delaying their first attack on England to 25 May 1917. This attack by *Kagohl* 3 did not reach London, nor did the second. But on 13 June the third attack did reach London, causing the most severe casualties of any bombing raid of the war. Together with the heavy casualties, the spectacle of Gothas leisurely bombing targets in London in broad daylight shocked the British public, with a significant ramification being the creation of the RAF from the RFC and RNAS on 1 April 1918 to provide a more coordinated and effective air defense capability. In the interim, the war cabinet agreed to double the size of both the RFC and RNAS as a direct result of these shocking raids.

Initial Gotha G.IV losses during the daylight attacks were light, but rapidly improving defenses caused a shift to night bombing starting the night of 3/4 September after eight daylight raids.

Above: Gotha G.IV 410/16 of *Kagohl* 3. The early Gotha G.IV bombers were painted light blue overall with natural metal engine cowlings.

The Gotha bombers were stable when fully loaded but only marginally stable when lightly loaded, as was normal during landing. Photos of loaded Gothas show bombs mounted under the front gunner's cockpit, far ahead of the center of gravity. Small, 12.5 kg bombs were also carried in bomb racks in the bombardier's cockpit. After release of these bombs the center of gravity moved aft significantly, making the Gothas much less stable in pitch. Accordingly, most operational losses were due to crashes during landing, when the lightly loaded bombers were least stable. And the night landings now required in its new role as night bomber emphasized the Gotha's handling problems during approach and landing; landing accidents causing 76% of all Gotha losses.

Below: Gotha G.IV 410/16 of *Kagohl* 3 after a crash during take-off. The plywood "Gotha Tunnel" that enabled the rear gunner to fire downward and to the rear is shown to advantage.

Gotha G.IV

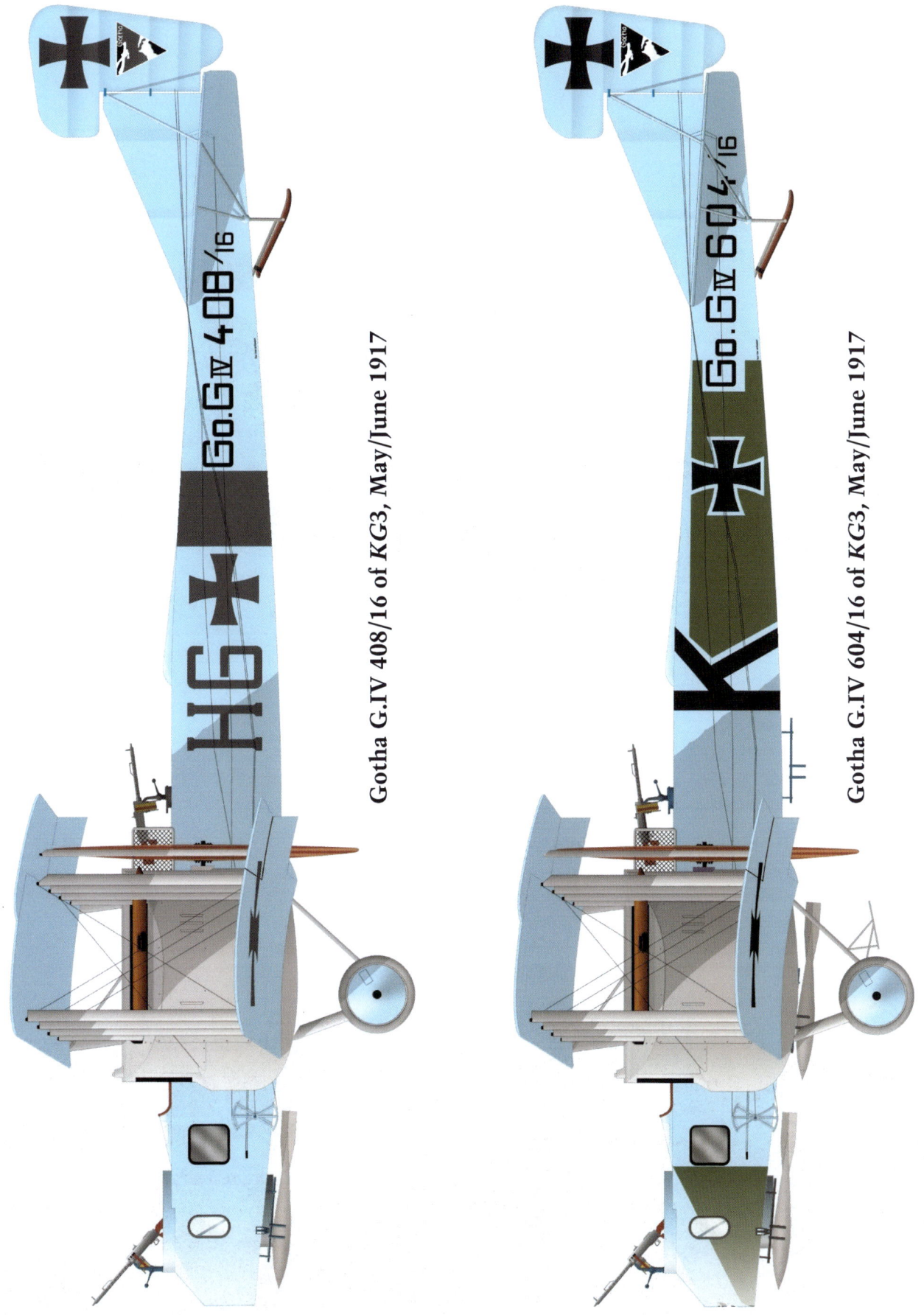

Gotha G.IV 408/16 of KG3, May/June 1917

Gotha G.IV 604/16 of KG3, May/June 1917

Above: A factory photograph of Gotha G.IV 607/16 carrying six 50-kg P.u.W. bombs beneath the fuselage. The far-forward carriage of the bombs under the nose illustrate the tail-heaviness of the basic design. The bombardier/forward gunner normally carried a number of 12.5 kg bombs in bomb racks inside his cockpit, which partially compensated for the center of gravity problems, at least until they were dropped.

Below: Gotha G.IV 602/16 crash-landed in neutral Holland at Sas van Gent on 28 September 1917 during a night raid on Britain. Another G.IV crashed and burned at Sneek the same day.

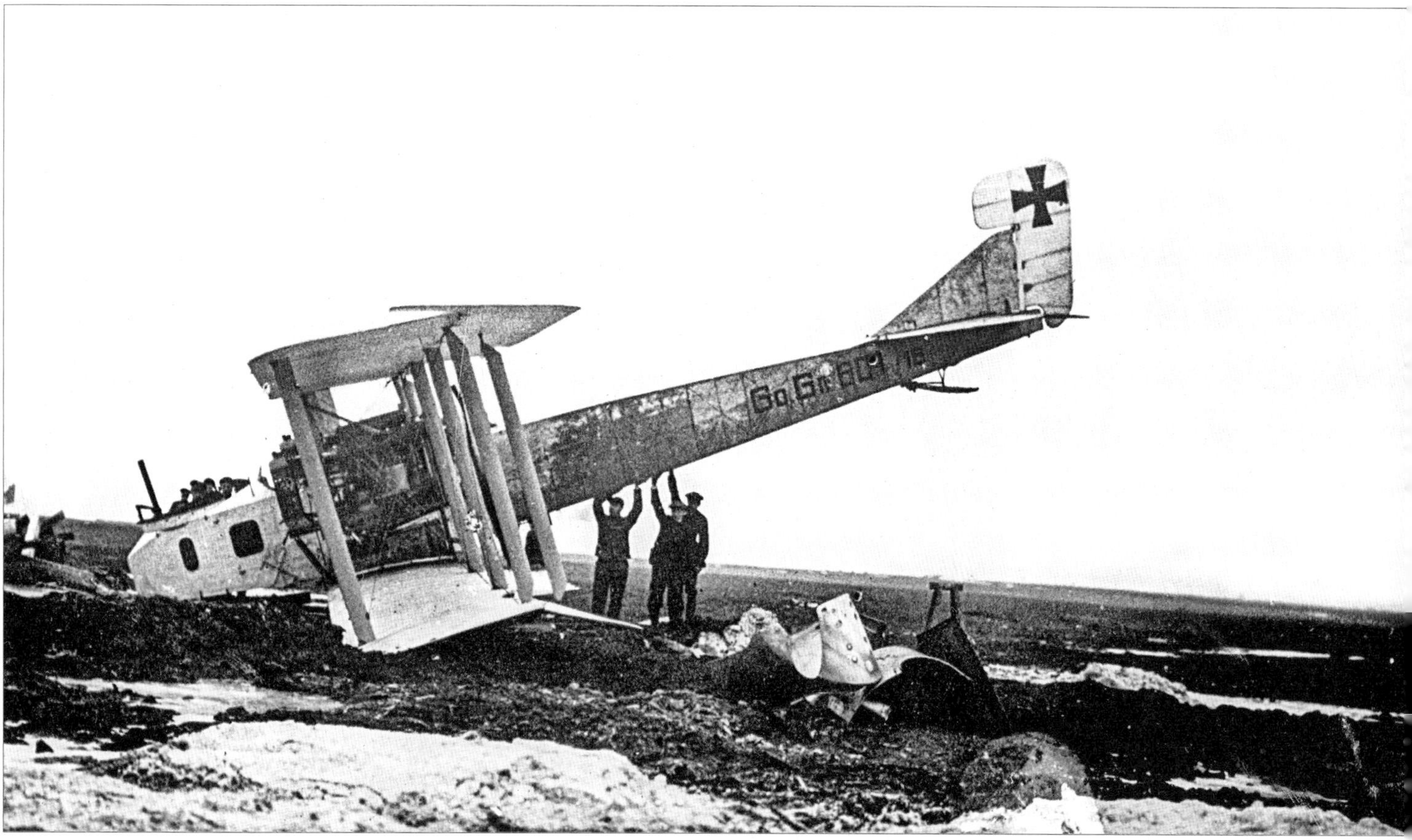

Above: A factory photograph of Gotha G.IV 601/16 after a landing accident. All aircraft design decisions are compromises between competing concerns, but landing accidents claimed 76% of the Gotha G.IV and G.V bombers lost to all causes. If the center of gravity is too far forward, the aircraft is excessively stable, too much download is required on the tail, which causes more drag, raising the nose to abort a landing becomes difficult, and nose-overs become more likely. However, the extreme number of Gotha landing accidents is proof that Burkhardt moved the center of gravity too far aft during development of the G.II, a mistake that could have been easily rectified, but was not.

Below: An unidentified Gotha G.IV after yet another landing accident.

Gotha G.IV

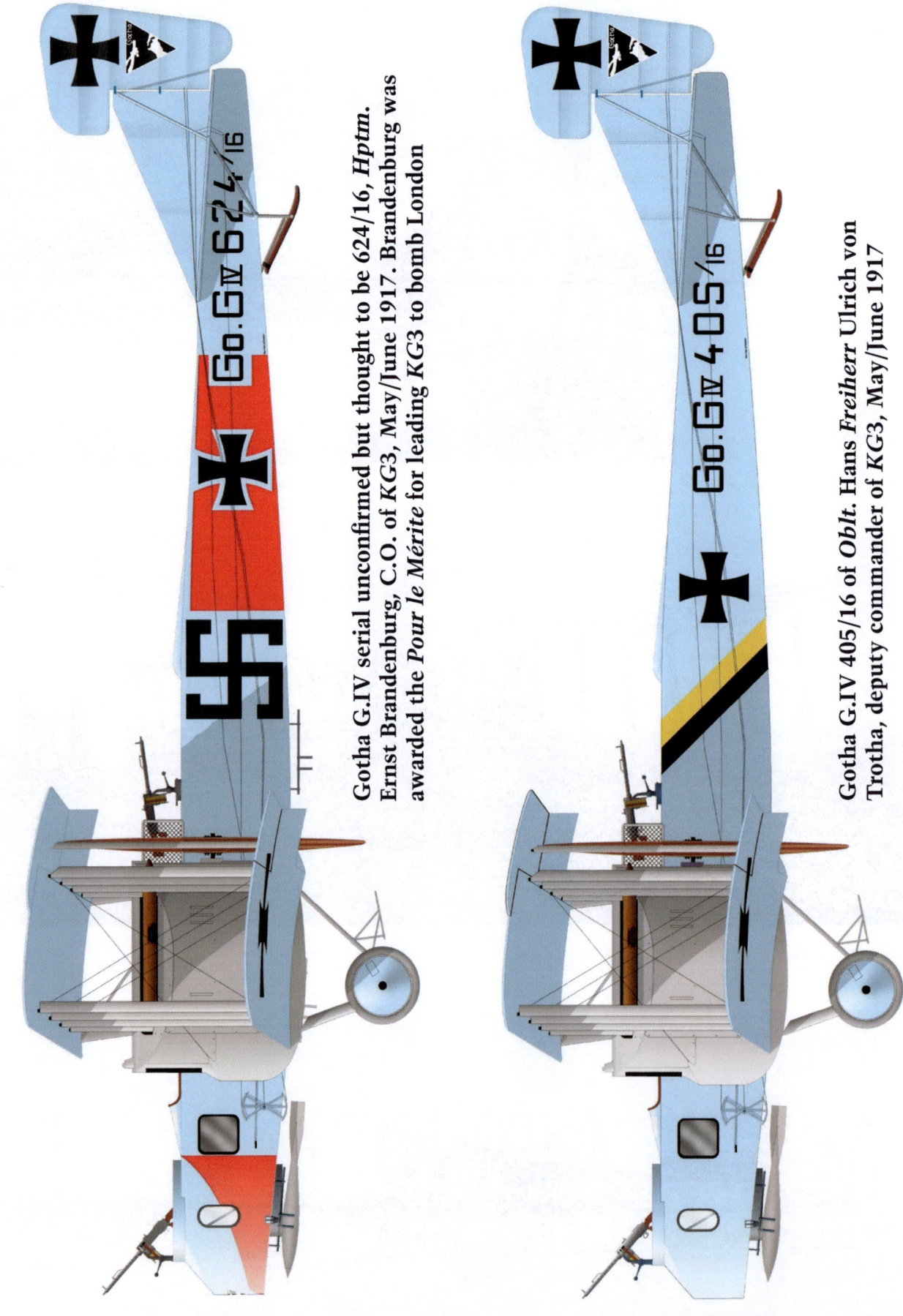

Gotha G.IV serial unconfirmed but thought to be 624/16, *Hptm.* Ernst Brandenburg, C.O. of *KG3*, May/June 1917. Brandenburg was awarded the *Pour le Mérite* for leading *KG3* to bomb London

Gotha G.IV 405/16 of *Oblt.* Hans *Freiherr* Ulrich von Trotha, deputy commander of *KG3*, May/June 1917

Above: A Gotha G.IV(LVG) in Austro-Hungarian service. The 39 Austro-Hungarian G.IV(LVG) bombers delivered were assigned Austrian serials 08.01–08.40; one had crashed in a "typical landing accident" by an LVG pilot and was not replaced. These aircraft had twin fuel tanks above the wing center section and were camouflaged in dark German hexagonal fabric for night bombing.

Below: Austro-Hungarian Gotha G.IV(LVG) 08.12 has drawn a crowd. The bomb racks are visible under the wing center section. Because German engine production was inadequate for German needs due to Allied numerical superiority and the Royal Navy's distant blockade, the Austrians had to provide their own 230 hp Hiero engines for their Gothas. The revised engine installation was a disaster; weak engine bearers and unsuitable propellers caused excessive engine vibration that led to piping leaks and failures of the structure and instruments. Despite great efforts by the maintenance staff, by late September 1918 the Austrian Gothas were virtually grounded and the aircrews reverted to the reliable, single-engine Brandenburg C.I for most bombing missions.

Gotha G.IV

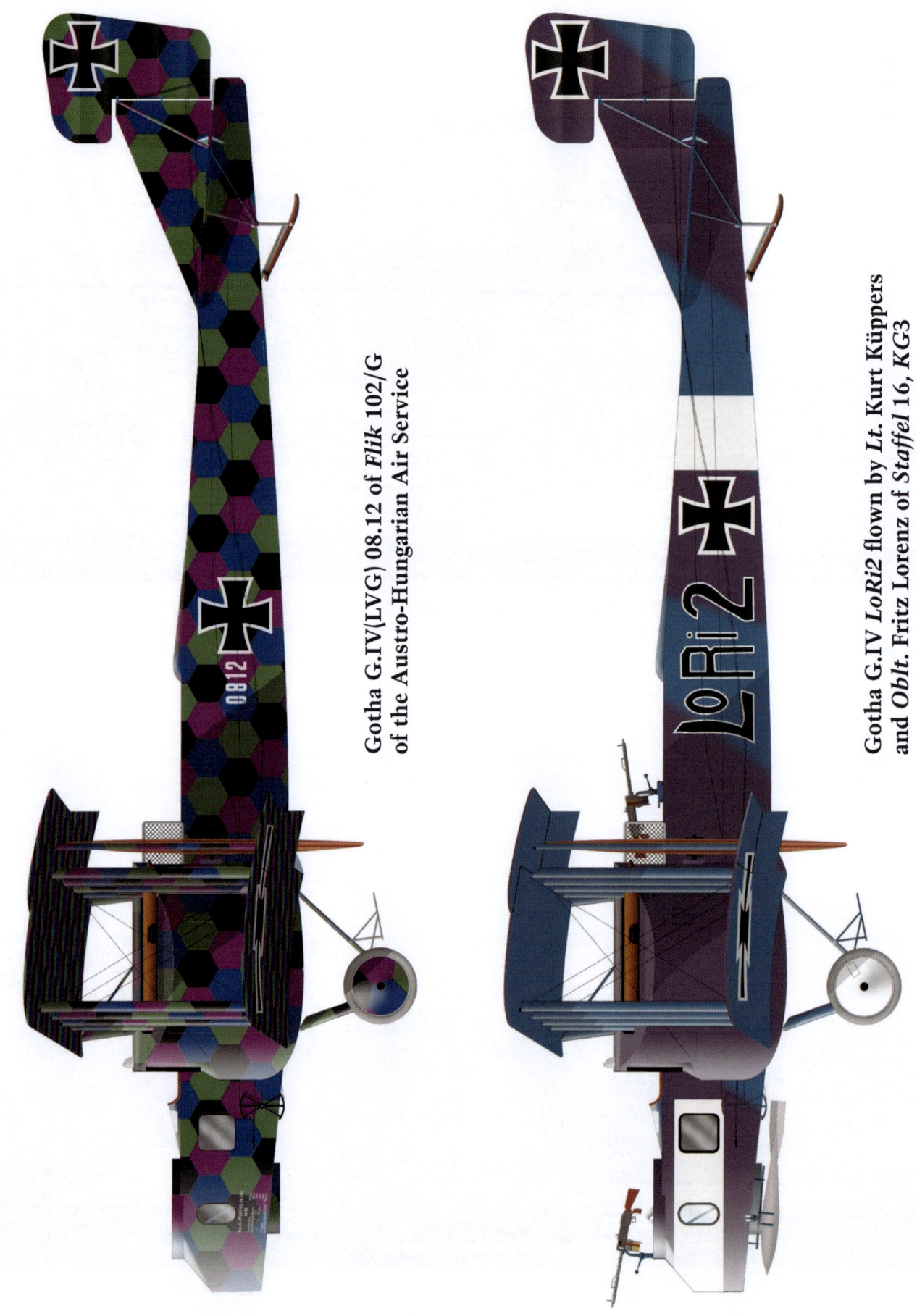

Gotha G.IV(LVG) 08.12 of *Flik* 102/G of the Austro-Hungarian Air Service

Gotha G.IV *LoRi2* flown by *Lt.* Kurt Küppers and *Oblt.* Fritz Lorenz of *Staffel* 16, KG3

Gotha G.V / G.Va / G.Vb

The Gotha G.V was a further refinement of the basic design. Experience had shown that housing the fuel tanks in the engine nacelles increased the chance of fire in the event of a crash. Therefore the fuel tanks of the G.V were moved to the fuselage away from the engines, which were enclosed in streamlined nacelles mounted between the wings. The engine remained the same 260 hp Mercedes D.IVa as used in the G.III and G.IV. Despite the nominal improvements incorporated into the G.V, the basic design had reached its peak. Furthermore, wartime shortages resulted in inferior materials being used to construct the G.V, making it heavier than the G.IV and reducing its performance. Additional equipment deemed necessary by operational experience aggravated the problem.

Regardless, *Idflieg* ordered 100 G.V bombers on 19 October 1916. The G.V passed its type test in July 1917 and the first aircraft reached *Bogohl* 3 in August 1917. The G.V arrived too late to participate in the daylight raids over the UK but were used for night bombing, performing both strategic raids over the UK and tactical bombing on the Western Front.

The Gotha G.Va was an attempt to improve flight safety. A re-designed 'box' tail with biplane horizontal stabilizers and elevators and twin fins and rudders improved engine-out controllability, even enabling turns into the running engine. A two-wheeled *Stossfahrgestell* mounted under the nose prevented nose-overs on landing to improve landing safety, the Gotha's Achilles Heel. The nose turret was also modified. Twenty-five of this transitional model were delivered.

The G.Vb was derived from the G.Va in a further attempt to improve flight safety. The 'box' tail was retained and, to reduce the pilot's workload and

Gotha G.V Specifications		
Engines:	2 x 260 hp Mercedes D.IVa	
Wing:	Span Upper	23.70 m
	Span Lower	21.90 m
	Area	89.5 m²
	Chord Upper	2.30 m
	Chord Lower	2.30 m
	Gap	2.22 m
	Sweepback	1.16°
General:	Length	12.40 m
	Height	4.30 m
	Empty Weight	2413 kg
	Loaded Weight	3648 kg
Maximum Speed:		140 km/h
Climb:	1000m	2.5 min
	2000m	8.5 min
	3000m	17 min
	4000m	29 min
Range:		840 km

Above: The distinguishing feature of the Gotha G.V was its new engine nacelles suspended between the wings. The fuel tanks were moved from the nacelles, their location in the G.IV, to the fuselage to reduce the danger of fire in a crash.

Above: This view of a Gotha G.V flown by *Kasta* 17 of *Bogohl* 3 shows its new engine nacelles, the night camouflage, and the original iron cross insignia over-painted to the new, straight-sided insignia. The bomber has an interesting marking of a constellation of eight-pointed stars with a line connecting them. Plates behind the wheels deflect debris from going through the propeller arc. A large load of bombs is carried beneath the wing center section.

improve maneuverability, most G.Vb bombers were equipped with Flettner servo controls on the upper ailerons. To further improve landing safety a four-wheeled undercarriage underneath each engine, similar to the original G.II prototype, was fitted. Unfortunately, the new undercarriage reduced the already compromised flight performance due to its greater weight and drag, and the Gotha G.Vb was inferior to the contemporary AEG G.V and Friedrichshafen G.IV based on comparative tests by *Idflieg*. Confirming this assessment, *Idflieg* placed an order with Gotha to build 50 Friedrichshafen G.IVa(Go) bombers under license, although the war ended before any were built. Although 80 Gotha G.Vb bombers were completed, the final few were delivered postwar directly to the Allies as part of the Armistice conditions.

Above: Front quarter view of an early Gotha G.V with faired struts supporting the engine nacelle.

Gotha G.IV & G.V

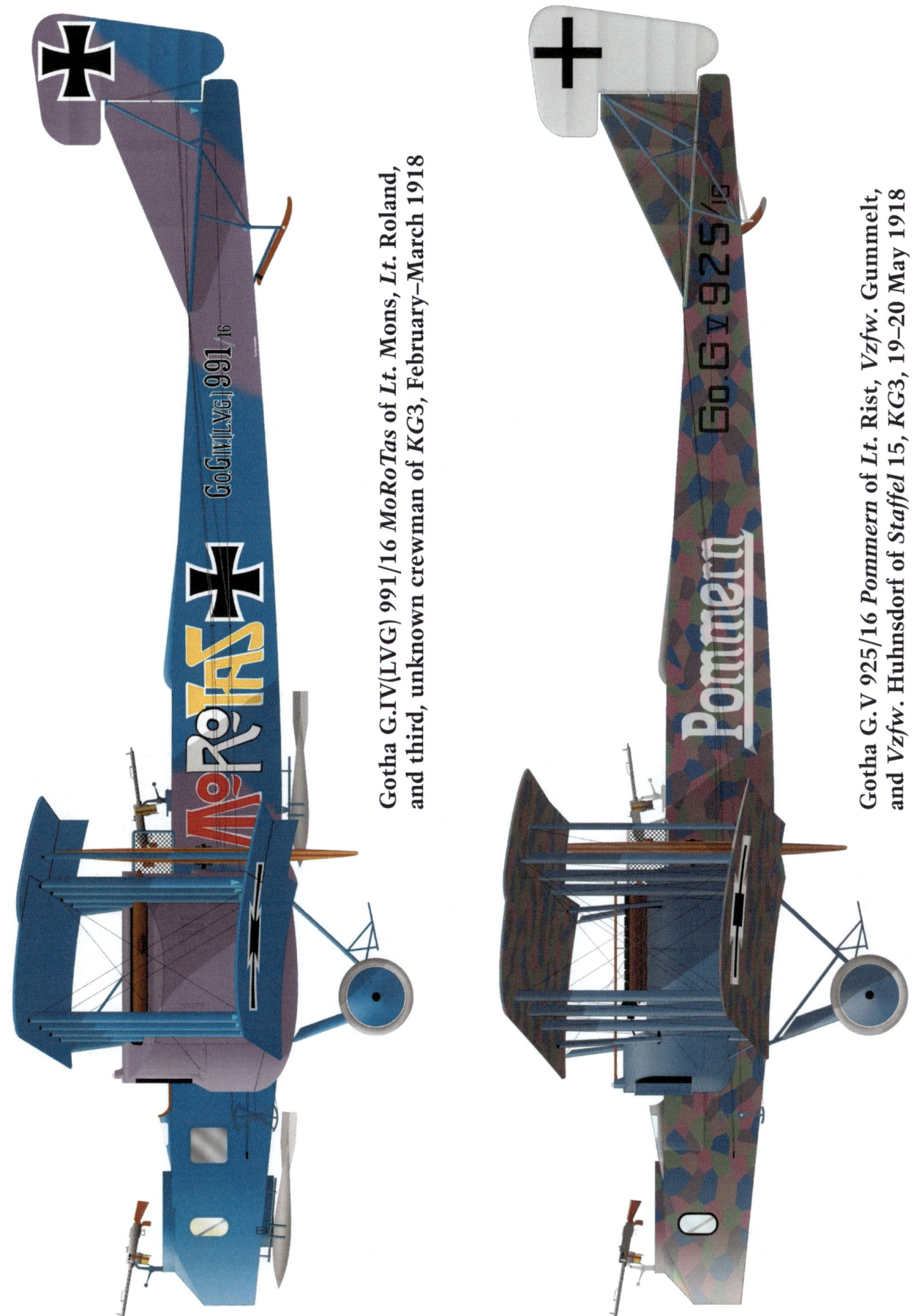

Gotha G.IV(LVG) 991/16 *MoRoTas* of *Lt.* Mons, *Lt.* Roland, and third, unknown crewman of *KG3*, February–March 1918

Gotha G.V 925/16 *Pommern* of *Lt.* Rist, *Vzfw.* Gummelt, and *Vzfw.* Huhnsdorf of *Staffel 15, KG3*, 19–20 May 1918

Gotha G.V

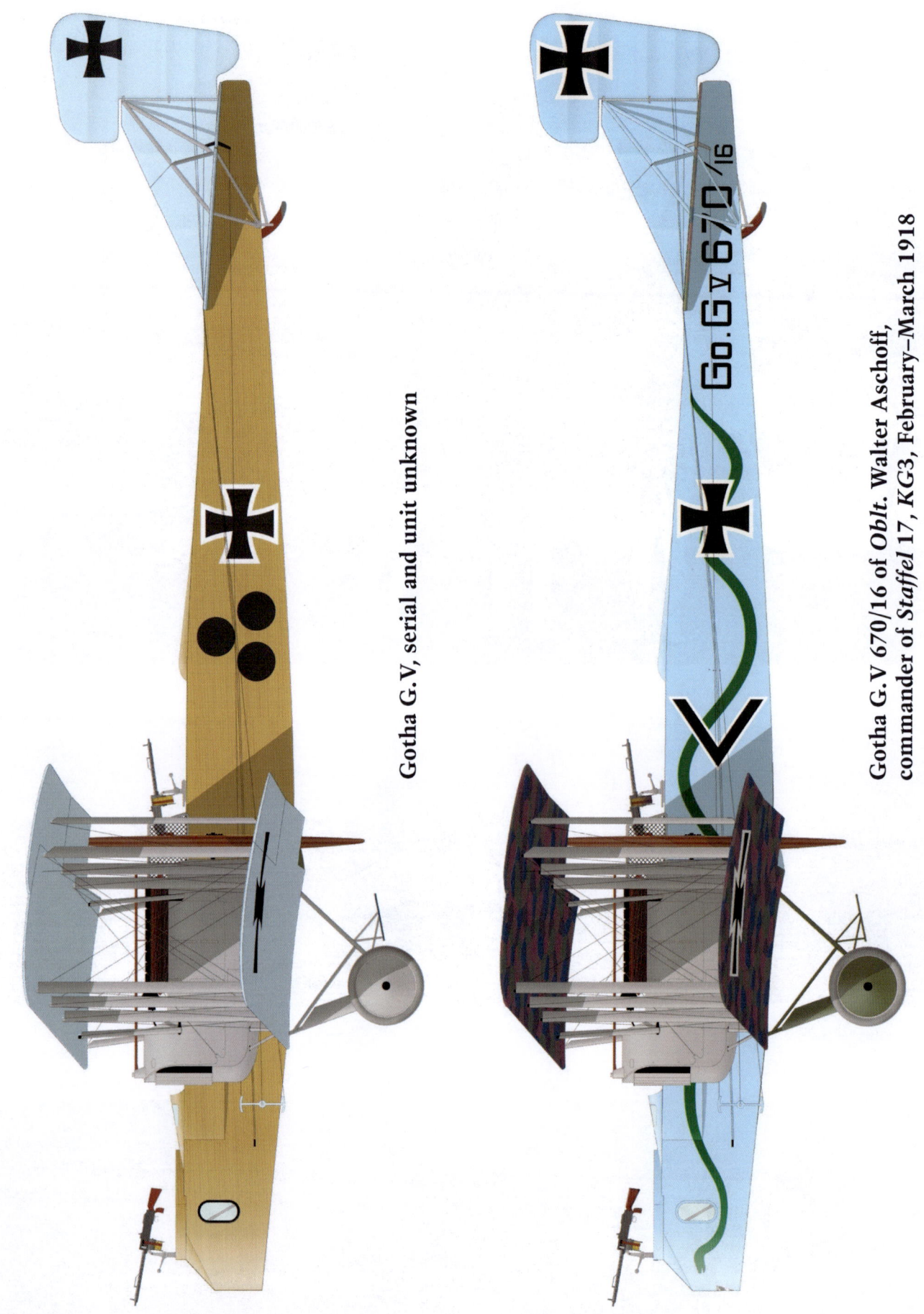

Gotha G.V, serial and unit unknown

Gotha G.V 670/16 of Oblt. Walter Aschoff, commander of *Staffel 17*, *KG3*, February–March 1918

Above & Below: Other than the change in location of the fuel tanks that affected the engine nacelles, the Gotha G.V was otherwise like the G.IV in all important respects. Not only did it offer no performance improvement, the deteriorating quality of the materials available for construction meant the G.V was a heavier airplane, reducing performance.

Above: Rear view of the Gotha G.V; this aircraft has only one over-wing gravity fuel tank.

Above: Some Gotha G.V bombers were fitted with an eight-wheel *Stossfahrgestell* undercarriage to prevent nose-overs.

Above: To improve controllability with one engine out, the Gotha G.Va with 'box' tail was developed. Above is the G.Va prototype fitted with an early version of the box tail.

Below: The Gotha G.Va prototype with the final configuration of the box tail used in production.

Gotha G.V

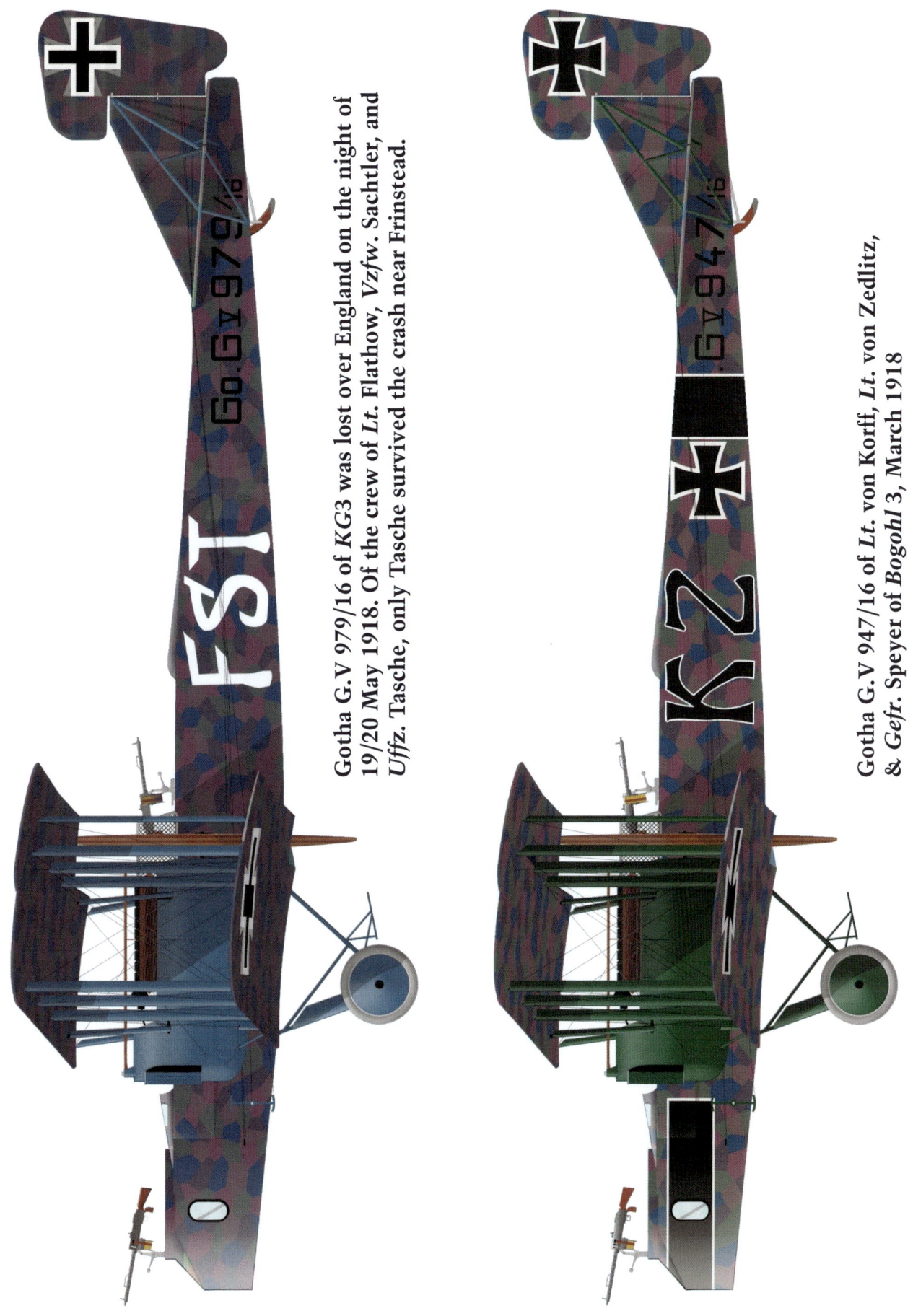

Gotha G.V 979/16 of *KG3* was lost over England on the night of 19/20 May 1918. Of the crew of *Lt.* Flathow, *Vzfw.* Sachtler, and *Uffz.* Tasche, only Tasche survived the crash near Frinstead.

Gotha G.V 947/16 of *Lt.* von Korff, *Lt.* von Zedlitz, & *Gefr.* Speyer of *Bogohl* 3, March 1918

Above: Gotha G.Va and G.Vb bombers stored in a hangar. The two G.Vb bombers in the foreground are not fitted with the Flettner servo-assist on the ailerons, normally a feature of the G.Vb.

Above: The tail of Gotha G.Va 723/17 brought down in France on 5 July 1918 wears a dramatic winged dragon personal marking. This aircraft is the one in the upper right of the photograph at the top of this page.

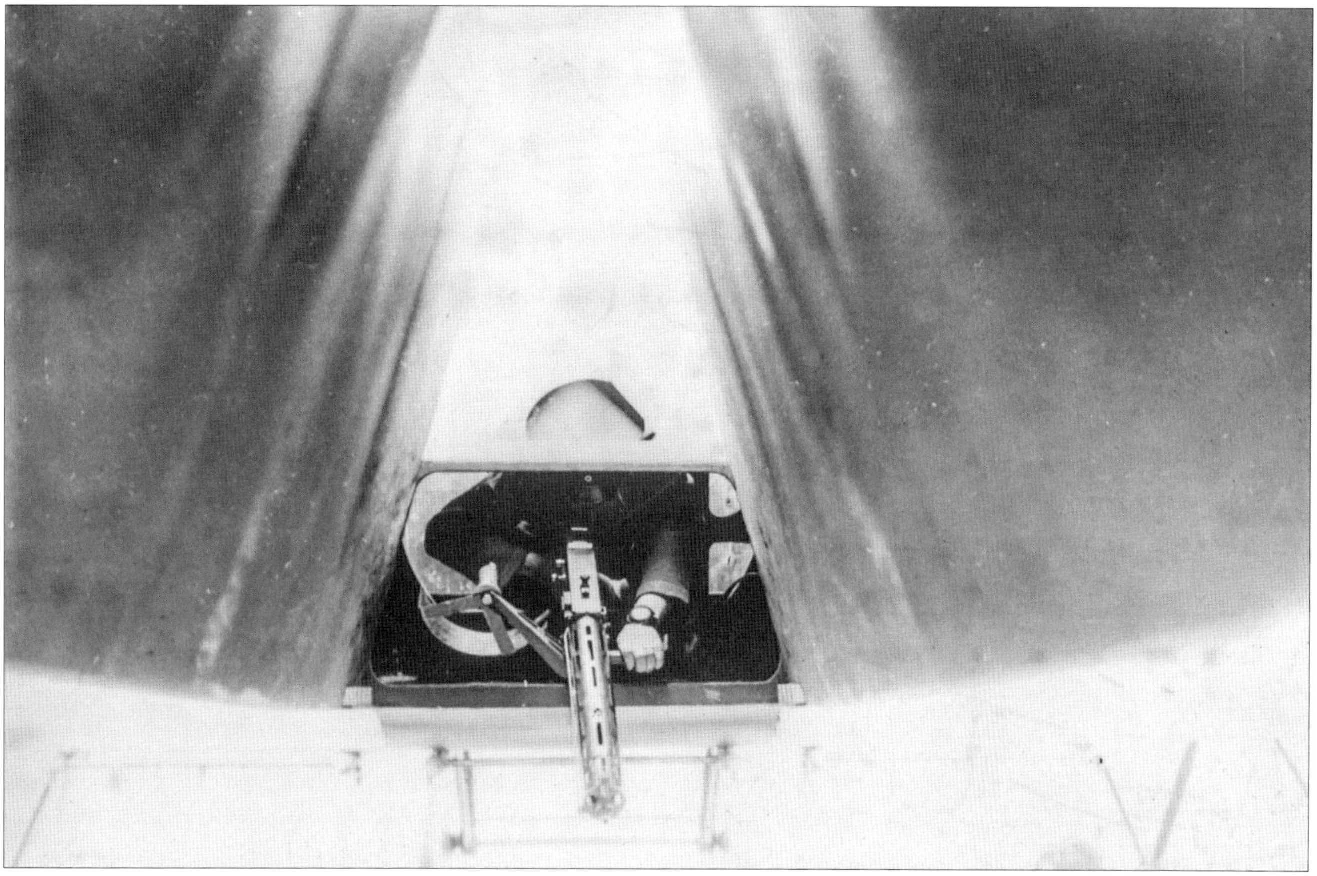

Above: The 'Gotha Tunnel' underneath the fuselage of a Gotha G.Va (notice the front undercarriage struts) as seen from the rear and being demonstrated by a gunner. The tunnel was introduced on the G.IV and carried forward into the G.V series; it enabled the rear gunner to protect the bomber against attacks from behind and below. A third machine gun could be mounted as shown, or the rear gunner's standard gun could fire downward through the slot in the rear fuselage as shown on page 134.

Left: Gotha G.V in night camouflage.

Above & Above Right: Two more views of the tail of Gotha G.Va 723/17 brought down in France on 5 July 1918.

Above & Below: The nose undercarriage indicates this is likely a Gotha G.Va. These well-known photos show a gunner demonstrating the field of fire of the front and rear gunner's positions in a night-camouflaged bomber. Below the gunner demonstrates how to protect against an attack from below by firing down through the tunnel in the fuselage. The screens on the sides of the rear gunner's cockpit keep his hands out of the propeller arcs. A rare Gotha G.VI prototype is in the background in the photo above.

Gotha G.Va & G.Vb

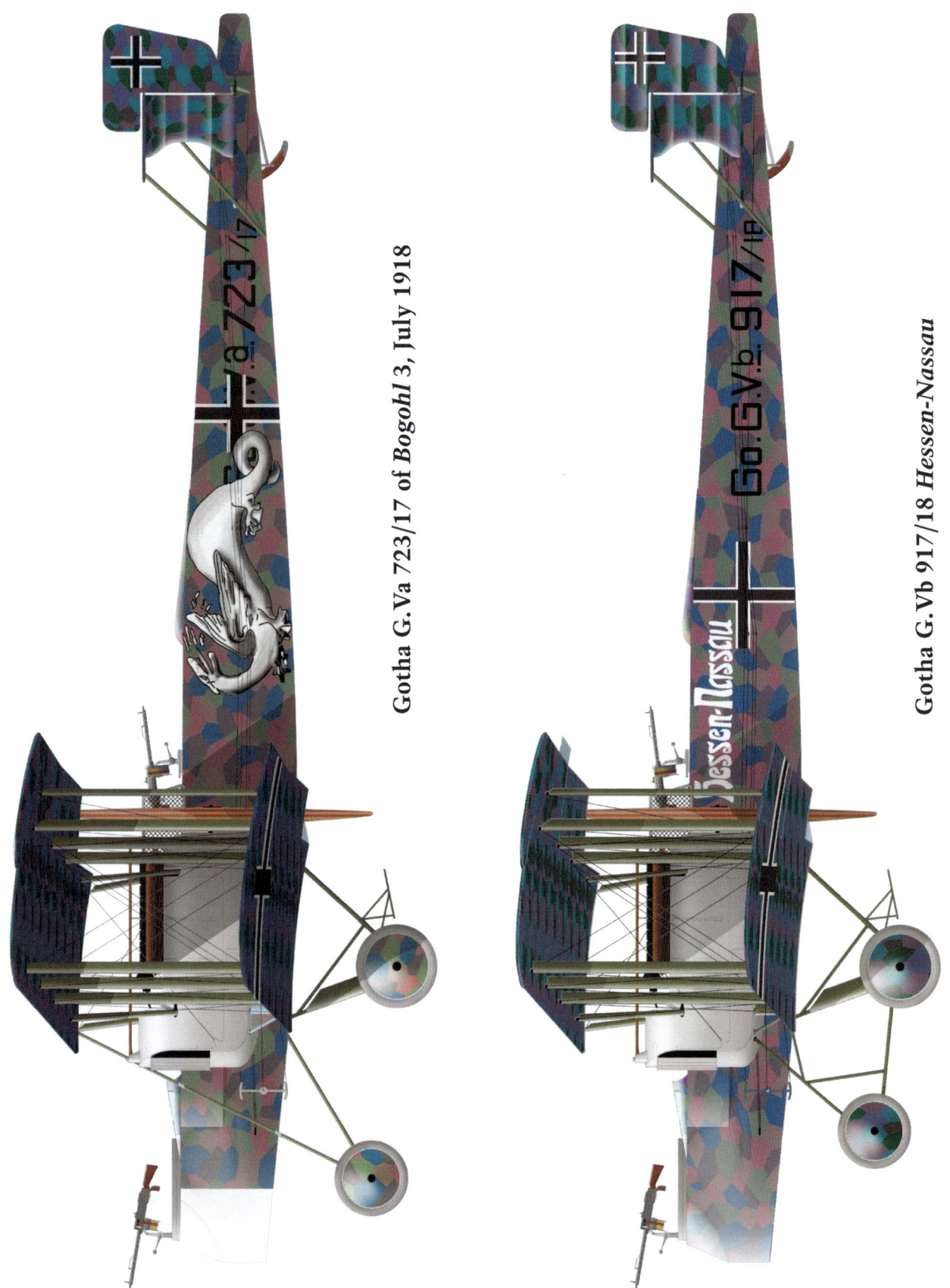

Gotha G.Va 723/17 of *Bogohl* 3, July 1918

Gotha G.Vb 917/18 *Hessen-Nassau*

Above & Below: Desperate to solve the landing accident dilemma that caused 76% of all Gotha bomber losses, the engineering team again modified the landing gear to create the Gotha G.Vb. The Gotha G.Vb reverted to an eight-wheel landing gear reminiscent of that fitted to the original G.II prototype. The additional forward-mounted wheels essentially eliminated the problem of nose-overs on landing, but at the expense of more weight and drag that reduced speed, range, and ceiling. This was a compromise the Gotha G.V could ill afford; it was already heavier than the G.IV due to inferior materials and more required equipment, and was now in an irreversible downward performance spiral.

Above: Gotha G.V *LoRi 3* of *Kasta* 16, *Bogohl* 3, and crew.

Below: Ground crewmen illustrate the range of bombs that Gotha bombers could carry with Gotha G.V 901/16 in the background. From left to right, the bombs are 12.5 kg, 50 kg, 100 kg, and 300 kg P.u.W. bombs. Reproduced as a Sanke card, this was a popular propaganda photograph.

Above & Below: Ground crewmen load bombs by hand onto Gotha G.V 901/16.

Above: Gotha G.V 901/16 fully loaded with bombs; five 50 kg and two 100 kg P.u.W. bombs are on the racks.

Below & Below Right: Groundcrew handling bombs to load on a Gotha. They are doing most of the work manually, with a primitive cart their only mechanical aid. These scenes could have been taken at any German base supporting twin-engine bombers in 1917 or 1918, regardless of the bomber's type.

Above: An impressive view of Gotha G.V 901/16 loaded and ready to start engines for its mission. The dark night camouflage shows the aircraft was normally used for night bombing, so the actual mission will not be flown until after dark. The Gothas ran most of their control wires outside the airframe, creating additional drag.

Above: Closeup of bombs hung under the extreme nose of a Gotha to provide a proper center of gravity. The Gothas were stable until the bombs were dropped, but pitch stability deteriorated significantly after that!

Above: The cockpit and flight controls of Gotha G.V 904/16 are typical for Gotha G.IV bombers as well. The cellon window above the instruments provided illumination during the day.

Above: Another view of a fully-loaded Gotha G.V. A number of 12.5 kg bombs were usually carried in the forward cockpit.

Below: With a Gotha G.V in the background, ground crewmen fill oxygen bottles with liquid oxygen for use by the bomber crew in flight. Without oxygen masks the crew sipped the oxygen in flight from a tube.

Gotha G.VI

Chief Engineer Burkhardt had been thinking about how to reduce drag to improve speed, and thought that eliminating one of the engine nacelles would accomplish that despite the resulting asymmetric configuration. Burkhardt decided the most efficient way to design the aircraft was a tractor engine in the fuselage and a pusher engine in a nacelle to starboard which also had an observer/gunner's cockpit in the front that extended forward of the tractor propeller for greatest field of fire.

To compensate for asymmetric thrust and drag, the starboard nacelle was placed closer to the centerline of the wing than the fuselage. The thrust lines of the engines were close together, reducing asymmetric thrust in case of an engine failure. Burkhard obtained German Patent number 300 676 for his innovative design on September 22, 1915. *Idflieg* approved further study on 7 September 1915 but held off construction until design studies were completed. On 26 June 1916 Gotha management approved construction of the G.VI, which was underway the next month, but *Idflieg* did not sign a contract until 5 July 1917, when three G.VI bombers, serial numbers G.370–372/17, were ordered.

Gotha G.VI Specifications		
Engines:	2 x 260 hp Mercedes D.IVa	
Wing:	Span Upper	23.70 m
	Area	89.5 m²
General:	Length	12.36 m

The first Gotha G.VI was flight-tested in the fall of 1917 and was badly damaged in a landing crash in November. The two-bay wing was similar to that used in the Gotha G.II and the fuselage was derived from the G.IV. The second Gotha G.VI had three-bay wings and large radiators. Flight tests of this aircraft began in January 1918 and continued through March. However, flight performance was mediocre and there were many radiator problems as well. In April work on the G.VI was abandoned to allow Gotha engineers to focus on the Gotha GL.VII and GL.VIII prototypes.

The Gotha G.VI is thought to be the first asymmetric aircraft design to fly, proving the asymmetric conception was practical. Although Burkhardt claimed that G.VI performance exceeded its conventional predecessors and that it flew very well, his description of its flying qualities may be optimistic given its crash-landing.

Right & Facing Page: The first Gotha G.VI prototype had two-bay wings. It is shown here after a landing accident in November 1917. The fuselage and tail were derived from the Gotha G.IV and the wings were similar to those of the Gotha G.II.

Below: The second Gotha G.VI prototype had larger, three-bay wings. Flight tests began in January 1918 and continued through March. Despite improved performance compared to the production G.V, in April work on the G.VI was abandoned due to mediocre flight performance. Problems with the radiators were a contributing factor. Furthermore, terminating G.VI development enabled Gotha engineers to focus on development of the more promising GL.VII and GL.VIII bombers.

Gotha GL Types

Realizing that the standard Gotha G series was too vulnerable to interception, *Idflieg* released a specification for a high-performance G-type, designated GL type (GL for lightened G-type). The specification required daylight operation at altitudes of 6–7,000 meters while carrying 300 kg of either bombs or photographic equipment.

The initial aircraft were designed by Rösner and Schleiffer and were completely different from the earlier Gotha bombers.

The Gotha GL ('GL' meant *erleichtertes-G flugzeug*, or lightened G aircraft) series were intended to meet the *Idflieg* requirement for long-range, high-speed aircraft for reconnaissance and bombing. In official records the designations were GL.VII, GL.VIII, and GL.IX, but the simple G-designation was used interchangeably. This has led to some confusion, but the G.VII and GL.VII were the same aircraft, as were the G.VIII and GL.VIII and the G.IX and GL.IX.

Gotha G.VII/GL.VII

Ordered on 15 March 1917, before daylight operations by Gotha G.IV aircraft had begun over the UK, four prototypes were built. These were designated Gs.1 by Gotha and G.VII and G.VIII by *Idflieg*. The first prototype, Gotha GL.VII 550/17, was completed in September 1917, the month that *Kagohl* 3 transitioned to night bombing of the UK due to strengthened British defenses and increasing losses of Gotha G.IV bombers on daylight missions.

The single-bay Gotha GL.VII 550/17, powered by two 260 hp Mercedes D.IVa engines, was one of four prototypes (G.550–553/17) ordered at the same time to explore different engine and wing configurations to determine the optimum layout to meet the demanding *Idflieg* requirements. A characteristic

Gotha G.VII/GL.VII Specifications	
Engines:	2 x 260 hp Mercedes D.IVa
Wing: Span	19.27 m
General: Length	9.60 m
Empty Weight	2420 kg
Loaded Weight	3140 kg
Maximum Speed:	180 km/h
Service Ceiling:	6000 m

of the design was the close spacing of the engines to minimize asymmetric thrust with one engine out. This configuration was based on a patent by *Ingenieur* Michael Schlieffer, a Gotha pilot, and used the venturi effect between the nacelle and fuselage to minimize drag and help counter asymmetric

Above: The prototype of a Gotha GL-type was the single-bay G.VII G.550/17, shown in September 1917. The engines rotated in opposite directions to minimize torque and the wheels are faired to reduce drag. However, the fuselage is suspended between the wings by a complex set of drag-producing struts.

Above: Gotha G.VII G.550/17 as built was very compact for a twin-engine aircraft.

Below: A Gotha GL-prototype, probably G.VII G.550/17 due to its wheel fairings, displays its very clean lines. Early GL-types had a single fin and rudder.

Above: Gotha G.VII G.550/17 shown in September–October 1917 with wheel fairings removed for flight tests.

thrust. The close engine spacing required a short nose that precluded a forward gunner.

The next prototype, Gotha G.VII 551/17, had a longer, two-bay wing cellule. The Gotha G.VII was faster than the Albatros D.III but, despite months of intensive testing to optimize engine, airframe, and propeller configurations to maximize speed, climb, and ceiling, the desired performance was not achieved. In late 1918 engines with experimental centrifugal superchargers were installed in this machine for testing.

The production standard GL.VII had a two-bay wing cellule with aerodynamically balanced ailerons on the upper wing and a box (biplane) tail. In 1918 eight aircraft were produced from an order for 55 aircraft (G.300–354/18). Gotha GL.VII 300/18 was the first of these and was powered by two Maybach Mb.IVa engines. It was sent to the front for evaluation in May 1918. GL.VII 302/18 went to the front on 27 June, followed by 301/18 on 1 August and 304/18 on 26 October. However, no further information is available on the evaluation of these aircraft and it is not known if they flew any operational sorties.

Owing to promising results obtained in early Gotha tests, both Aviatik and LVG were awarded contracts in February 1918 to build the Gotha G.VII under license. Aviatik built 30 G.VII aircraft (G.100–129/18), few of which were handed over to the Allies after the war. Aviatik-built Gotha G.VII(Av) 112/18 was experimentally fitted with nose radiators for flight trials.

Below: One of the Gotha GL-prototypes, still with a single fin and rudder, now has the fuselage faired into the lower wings to reduce drag.

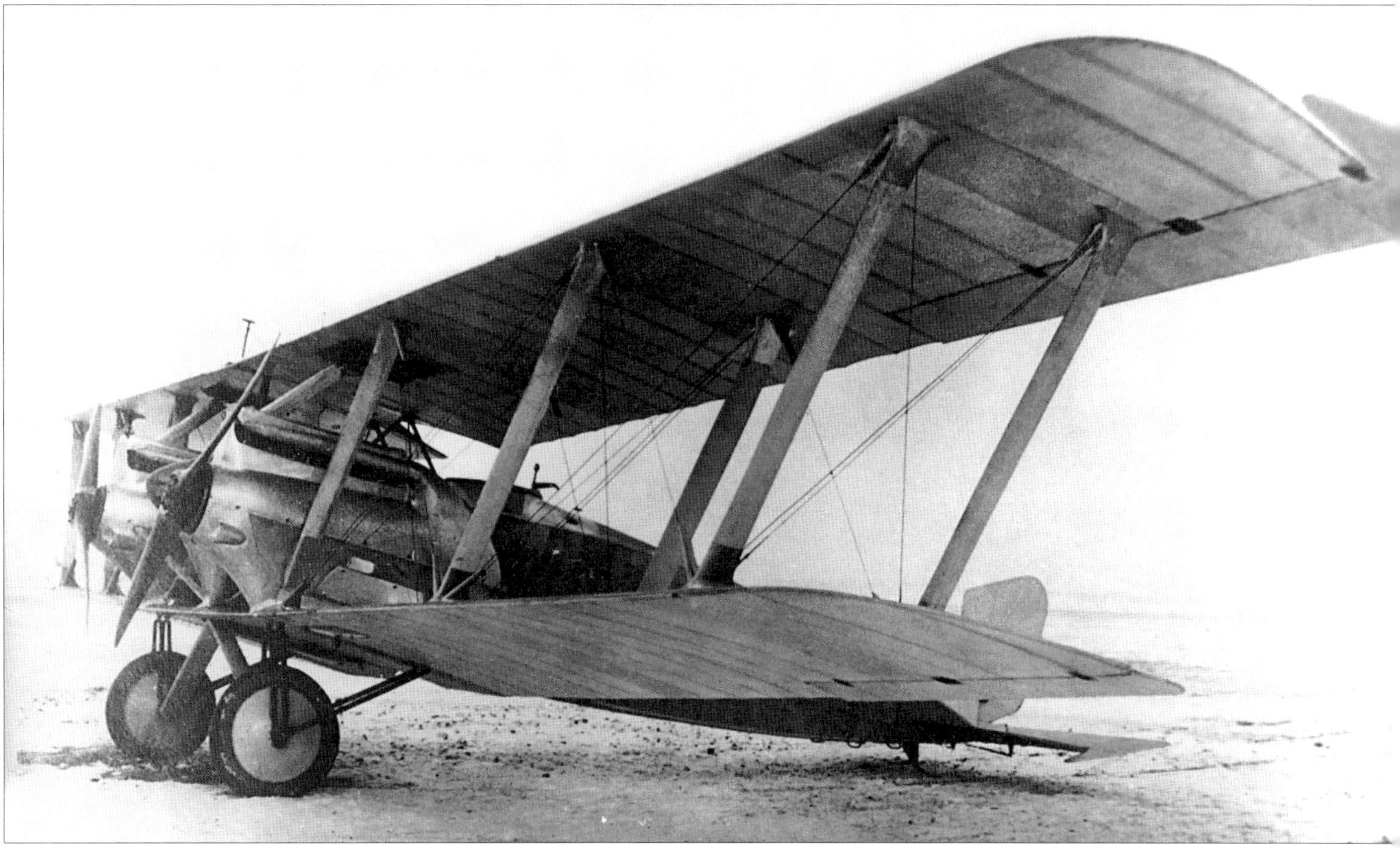

Above: This Gotha GL-prototype may be the same aircraft as the one at the bottom of the previous page, but the propeller spinners have been removed.

Below: A Gotha GL.VII G.300/18 represents the production version of the aircraft. Now it has a 'box' tail for better directional control with one engine out, no wheel fairings, and no propeller spinners. The wings are now longer, with two-bays of struts, and the aileron aerodynamic balances have slots in them.

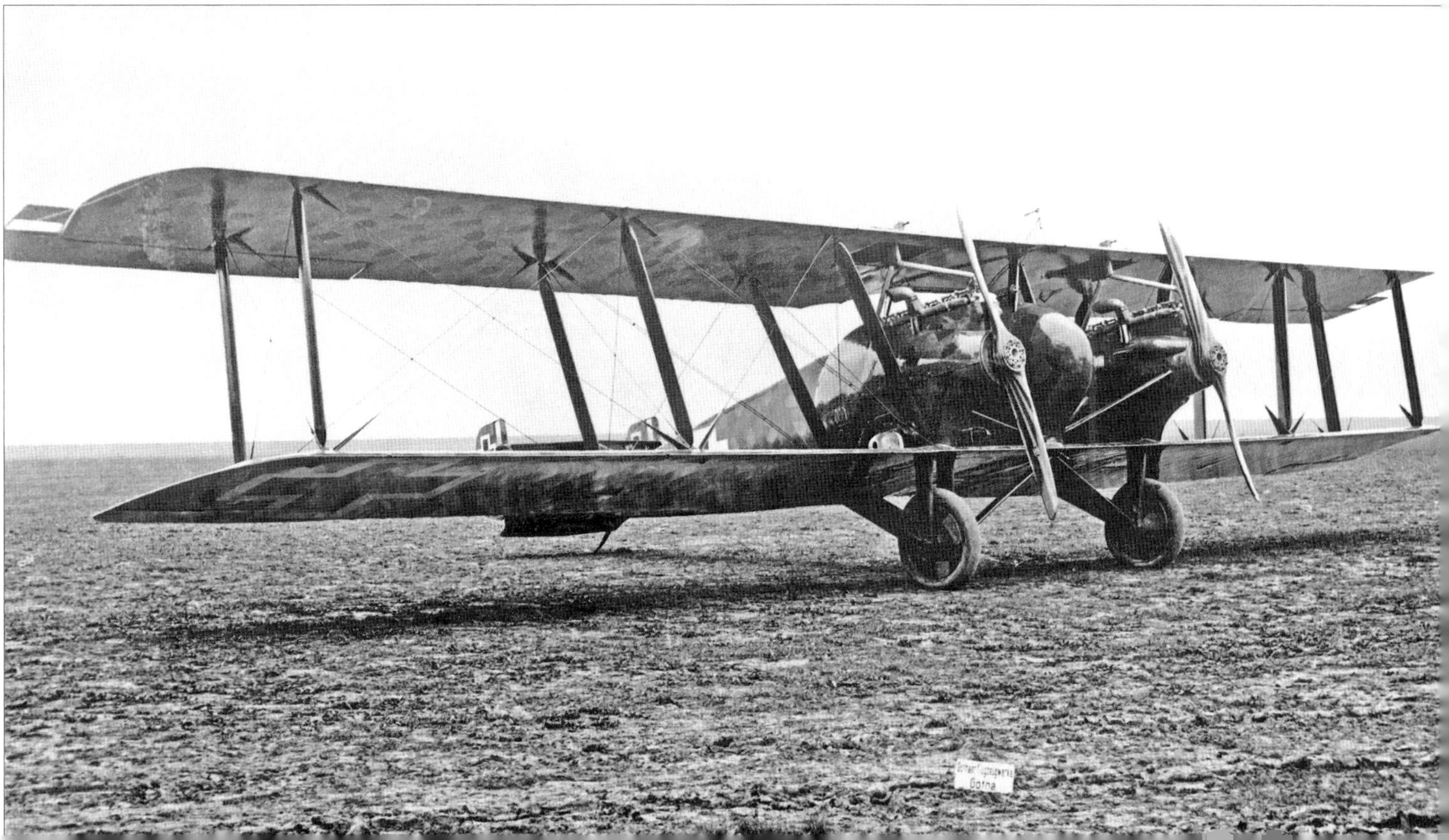

Above: A rear quarter view of Gotha GL.VII G.300/18, the production version of the GL.VII. The 'box' tail is visible and the rudders, like the ailerons, have aerodynamic balances with slots in them.

Below: This Gotha GL.VII in the factory has no markings, but the wings are covered in regular camouflage fabric with hexagons. The fuselage has been spray-painted in hexagons with soft edges. This may be one of the GL-prototypes. The 'box' tail has the more common aerodynamic balances without slots in them.

Above: Aviatik-built Gotha GL.VII(Av) G.112/18 was fitted with nose radiators for evaluation. The aileron aerodynamic balances do not appear to have slots. Aviatik built 30 GL.VII bombers.

Below: This Gotha GL.VII has an aluminum nose and propeller spinners. The radiators were mounted in the upper wing above the engines, the position used for production aircraft.

Gotha G.VIII/GL.VIII

Intended for high-speed day bombing, the Gotha GL.VIII was fitted with an enlarged, three-bay wing to improve climb and ceiling. One fixed, forward-firing machine gun was fitted for the pilot, and the gunner had both dorsal and ventral flexible guns for defense.

The Gotha GL.VIII prototype, G.VIII 552/17, was powered by Mercedes D.IVa engines, but the production standard was the over-compressed 260 hp Maybach Mb.IVa for more power at high altitude. All control surfaces were aerodynamically balanced and the biplane tail was retained for improved control with one engine out. The PuW bombs were carried externally under the fuselage.

Gotha GL.VIII 307/18, powered by two 260 hp Maybach Mb.IVa engines, was sent to the Döberitz test center on 19 October 1918. The upper ailerons were fitted with Flettner servo controls for lighter control forces. According to an *Idflieg* appreciation, the performance of these aircraft "lagged far behind expectations." On the other hand, the war ended before sufficient experience was gathered to evaluate the aircraft adequately under combat conditions.

Gotha G.VIII/GL.VIII Specifications		
Engines:		2 x 260 hp Maybach Mb.IVa
Wing:	Span	21.73 m
	Chord	2.05 m
	Gap	2.00 m
General:	Length	9.79 m
	Useful Load	1030 kg
	Loaded Weight	3706 kg
Maximum Speed:		180 km/h
Service Ceiling:		6000 m

Below: Although some Gotha GL prototypes appeared to carry a fixed gun for the pilot, the Gotha GL.VIII high-speed day bomber was the only type for which a fixed machine gun for the pilot was specified. Here the gunner demonstrates the field of fire of his dorsal gun; the fixed pilot's gun is also clearly visible. The gunner also had a flexible ventral gun to defend against attacks from below. The lozenge printed fabric and clean lines are evident.

Above & Below: Gotha GL.VIII G.307/18, a production GL.VIII, was powered by two 260 hp over-compressed Maybach Mb.IVa engines optimized for power output at high-altitudes for maximum ceiling and speed at altitude. The slanting struts supporting the upper wingtips characterize the production GL.VIII and differentiate it from the GL.VII.

Gotha GL.VIII 307/18

Gotha G.IX/GL.IX

In February 1918 LVG was given a production contract to build 30 Gotha GL.VII reconnaissance airplanes (G.200–229/18) and 70 Gotha G.IX bombers (G.230–299/18). The 260 hp Maybach Mb.IVa was the normal power plant. Apparently all of the GL.VII(LVG) aircraft were built but no photographs of this version have been found. The G.IX day bomber, given an increased wing-span and length, was equipped to carry five 50 kg P.u.W. bombs under the fuselage. According to a French inspection report, some 96 Gotha G.IX bombers were stored at the LVG factory on 15 December 1918. After the armistice a number of these bombers were given to the Allies. For example, Belgium received 23 G.IX bombers that were flown operationally by the Belgian air service. A Gotha G.IX(LVG) 289/18, delivered to Japan post-war as reparations, was powered by two 260 hp Mercedes D.IVa engines.

Gotha G.IX/GL.IX Specifications	
Engines:	2 x 240 hp Maybach Mb.IVa

Above: Gotha G.IX(LVG) G.257/18 is covered in dark, printed camouflaged fabric.
Below: Gotha G.IX(LVG) G.263/18 displays the LVG-style markings and is armed with a P.u.W. bomb under the fuselage.

Above: This Gotha G.IX(LVG) illustrates the extended wing-span of this version of the GL series. A British crew attempted to fly this aircraft back to Britain, but it crashed when leaving Bickendorf and the crew was killed.

Below: This Gotha G.IX(LVG) illustrates the type's clean lines. It is serving in Belgium postwar. The number '250' on the nose *may* indicate it is serial G.250/18, but there is no confirmation. Roundels are being painted under the wings but are not yet complete. The aircraft is covered in dark, printed camouflage fabric.

Above & Below: Gotha G.IX(LVG) in Belgian service postwar. These views illustrate details and emphasize the aerodynamic lines of the GL series. Problems of propeller selection for maximum ceiling and speed at altitude, always a challenge with fixed-pitch propellers, prolonged development of the GL-series.

Above: Gotha G.IX(LVG) in Belgian service postwar.

Below: The Gotha G.IX(LVG) in this postwar view has no visible markings other than the Belgian colors on the rudders, but its hexagonal camouflage fabric is well illustrated. The aileron aerodynamic balances have slots, but the aerodynamic balances on the rudders do not. A DH.9 is visible in the background.

Above: This Gotha G.IX(LVG) 299/18 illustrates its LVG-style markings, its camouflage fabric, and the fairings over the radiators in the upper wing above the engines.

Below: Gotha G.IX(LVG) serving in Belgium postwar illustrates the type's clean lines. It appears to be on public display. A Friedrichshafen bomber is in the background.

Gotha G.X/GL.X

On 16 August 1918, *Idflieg* ordered three Gotha GL.X prototypes (G.1725–1727/18), each powered by two 185 hp BMW.IIIa engines. Designed by Burkhardt, two GL.X versions were envisioned, one a fast, high-altitude reconnaissance machine and the second an armored ground-attack aircraft fitted with a single, forward-firing 20 mm Becker cannon and five fixed machine guns. A gunner had a sixth, flexible machine gun for rear defense. The first two-bay GL.X, a more compact design than the G.IX bomber, was reported ready for flight testing on 1 January 1919, subject to clearance from Gotha management.

It is difficult to understand how the same engine and airframe combination was expected to perform well at both extreme high and low altitudes.

Gotha G.X/GL.X Specifications	
Engines:	2 x 185 hp BMW.IIIa

Above: The Gotha G.X prototype had two-bay wings and was powered by two 185 hp BMW.IIIa engines. Somehow this airframe and engine combination was supposed to excel as both a high-altitude reconnaissance aircraft and a low-level ground-attack aircraft, seemingly contradictory requirements. The massive WD27 looms in the background.

Above: Gotha GL fuselages at the USAS center at Romorantin postwar show the fuselage cross section.

Left: Gotha GL.IX in Belgium postwar.

Gotha G-Type Bomber Production Summary

Order Date	Type	Qty	Serials	Notes
April 1, 1915	G.I	6	9–14/15	Plus one Friedel-Ursinus prototype, B.1092/14. Production G.Is delivered 27 July–8 Sep. 1915.
July 15, 1915	G.I	6	40–45/15	Delivered 22 Sep.–5 Nov. 1915.
Oct. 10, 1915	G.I	6	100–105/15	Delivered 24 Jan.–20 March 1916.
Dec. 18, 1915	G.II	10	200–209/16	Delivered Aug.–Sep. 1916
May 3, 1916	G.III	25	376–399/16	Delivered 16 Oct. 1916–26 March 1917
Aug. 4, 1916	G.IV	12	401–412/16	
Oct. 19, 1916	G.IV	25	600–624/16	
Nov. 23, 1916	G.IV	15	649–663/16	
Dec. 4, 1916	G.IV(LVG)	50	980–1029/16	
Dec. 18, 1916	G.IV(SSW)	40	1055–1094/16	Delivered July 1917–February 1918
May 1917	G.IV(SSW)	40	200–239/17	Delivered Dec. 1917–Aug. 1918
Aug. 1917	G.IV(LVG)	50	100–149/17	
Oct. 19, 1916	G.V	20	664–683/16	Delivered Aug. 1917–April 1918
Oct. 19, 1916	G.V	80	900–979/16	Delivered Aug. 1917–March 1918
Oct. 15, 1917	G.Va	25	700–724/17	Delivered 3 Apr.–22 May 1918
Oct. 15, 1917	G.Vb	25	725–749/17	Delivered 4 June–15 Aug. 1918
May 18, 1918	G.Vb	25	913–927/18	Delivered 30 June–6 Sep. 1918
July 26, 1918	G.Vb	15	1450–1464/18	Delivered 9 Sep.–2 Oct. 1918
Aug. 4, 1918	G.Vb	15	1778–1792/18	Delivered 18 Oct.–14 Dec. 1918
July 5, 1917	G.VI	3	370–372/17	Only first two built; third cancelled.

In addition to these aircraft delivered to *Idflieg* orders, LVG built an additional 40 G.IV bombers that were delivered to an Austro-Hungarian order and used 230 Hiero engines built in Austria. The squadrons received these aircraft, Austro-Hungarian serials 08.01–08.40, in March–April 1918.

Gotha GL-Type Production Summary

Manufacturer	Type	Qty	Serials	Notes
Gotha	G.VII & G.VIII	4	550/17 – 553/17	Prototypes; all 4 completed.
Gotha	G.VII & G.VIII	55	300/18 – 354/18	Production aircraft; only 8 were completed.
Aviatik	G.VII(Av)	100	100/18 – 129/18*	At least 30 completed.
LVG	G.VII(LVG)	30	200/18 – 229/18	Reconnaissance aircraft. All 30 believed to have been completed.
LVG	G.IX(LVG)	70	230/17 – 299/18	Bombers. Unknown number, perhaps all, completed.
Gotha	G.X	3	1725/18 – 1727/18	Prototypes. At least one completed.

Notes: 1. *Rest of serials unknown.
2. According to a French inspection report 96 Gotha G.IX aircraft were stored at the LVG factory on 15 Dec. 1918. Many of these were transferred to the Allies postwar, especially Belgium.

Gotha GL.VII & GL.IX

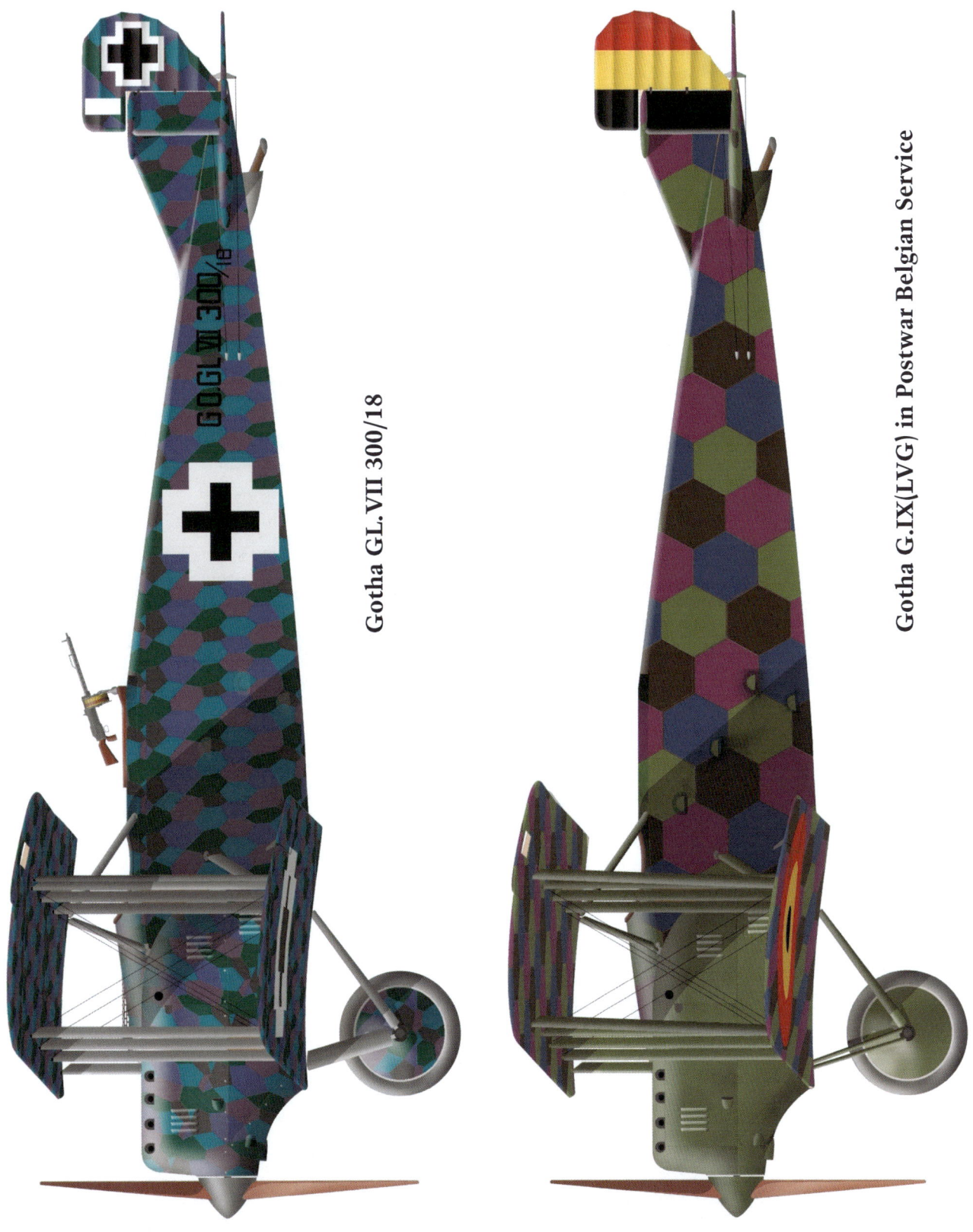

Gotha GL.VII 300/18

Gotha G.IX(LVG) in Postwar Belgian Service

Halberstadt G.I

Halberstadt only built one twin-engine aircraft during the war, the sole G.I prototype. The G.I was powered by two 160 hp Mercedes D.III engines. The aircraft was of conventional design and construction. The engines were mounted between the wings in tractor configuration. The biplane wings had two bays of bracing outboard of the engines and the tail had three rudders; only the central rudder had a fixed vertical fin.

A crew of three was carried; the pilot sat in the middle cockpit and there were gunners fore and aft, each with a flexible machine gun. A bomb load of 200 kg could be carried.

Halberstadt G.I Specifications		
Engines:	2 x 160 hp Mercedes D.III	
Wing:	Span	15.50 m
General:	Length	9.00 m
	Height	3.2 m
	Empty Weight	1,220 kg
	Loaded Weight	1,895 kg
Maximum Speed:		152 km/h
Climb:	1,000 m	7 min.
Duration:		4 hrs.

A single prototype was built during the winter of 1915/1916 but no series production was undertaken.

Above & Right: The sole prototype of the Halberstadt G.I is shown with its 160 hp Mercedes D.III engines running. The G.I was a relatively clean design with frontal radiators, nicely-cowled engines, and simple landing gear. It was compact for a twin-engine bomber and faster than most competitors, but was not selected for production; perhaps its bomb load was too small.

Hansa-Brandenburg G.I

The story of the Hansa und Brandenburgische Flugzeugwerke AG is too complex to detail here. Hansa-Brandenburg, commonly known as Brandenburg, had production facilities in both Germany and the Austro-Hungarian Empire, but aircraft design remained at Briest in Germany. As a result of unsatisfactory testing of early Brandenburg aircraft, the German Army would not consider future aircraft from Brandenburg and the company focused on seaplanes for the German Navy and both land planes and seaplanes for Austria-Hungary. In addition to designing and building at least 15 prototypes, Brandenburg also built 260 production aircraft for Austria-Hungary.

The path to the Brandenburg G.I began when the *LFT* (*Luftfahrtruppen*, the Austro-Hungarian Air Service) ordered a battleplane from Brandenburg on May 25, 1915. The battleplane concept was embraced to some extent by all combatants and was the result of thinking of air combat as analogous to naval combat. Actual air combat soon showed this was an outmoded concept and some aircraft designed as battle planes became bombers. In any case, the prototype aircraft designed to meet these requirements, the Brandenburg 05.05, was powered by two 160 hp Mercedes engines in tractor configuration. The 05.05 was a conventional biplane utilizing the typical wood, wire, and fabric construction of the era with the fuselage covered with plywood. For directional control with an

Brandenburg G.I Specifications		
Engines:	2 x 160 hp Daimler	
Wing:	Span	18.00 m
	Chord	2.00 m
	Gap	2.10 m
	Area	70.0 m²
General:	Length	9.80 m
	Height	3.6 m
	Empty Weight	1,776 kg
	Loaded Weight	2,740 kg
Maximum Speed:		144 kmh
Climb:	1,000m	8 min

Left & Below: The sole Brandenburg 05.05 was powered by two 160 hp Mercedes engines. It was a conventional wood, wire, and fabric design of the era.

Above: The sole Brandenburg 05.08 was powered by two 160 hp Mercedes engines. It was an intermediate step between the 05.05 prototype and the production Brandenburg G.I bomber, which was very similar to the 05.08.

Left: The pilot of the Brandenburg 05.08 sat well aft and shared his cockpit with the rear gunner. The G.I adopted this configuration but when later modified the pilot's cockpit was moved forward for better field of view.

inoperative engine it had three rudders, only the central of which also had a fixed fin. Fore and aft gunners each had a flexible machine gun.

After delivery to the *LFT* the 05.05 was sent to the Isonzo Front for operational evaluation, but no details are available. Damaged in a storm on October 15, 1915, the 05.05 was repaired and bomb racks were installed. The modified aircraft was sent to *Flik* (*Fliegerkompagnie* – aviation company) 19 in March and was destroyed on April 6, 1916, in a landing accident while returning from a bombing raid.

On July 21, 1915, the *LFT* ordered an improved aircraft which resulted in the 05.08 prototype. The 05.08 had a robust steel-tube aircraft instead of the wood frame of the 05.05. Although basically similar to the earlier aircraft, the detail design was different to give more strength. Again power came from two 160 hp Mercedes engines. In May 1916 the 05.08 was assigned to *Flik* 19 for operational evaluation and participated in several bombing raids.

Encouraged by the successful but brief trials of the 05.05 and the 05.08 at *Flik* 19, the *LFT* ordered 72 production G.I bombers based closely on the 05.08. But by April 1916, *Flars* (*Fliegerarsenal* – the Austro-Hungarian aviation arsenal) reported that "twin-engined aircraft were unfit for operational service because of their inability to remain aloft on one engine, making long-range missions impossible." Orders for the G.I were therefore reduced to 39 aircraft.

Unfortunately, the structural strength and flight characteristics of the G.I were marginal, and after Hansa-Brandenburg refused to fix the problems at no charge, the aircraft caused a serious argument between Brandenburg and the *LFT*. Because the company was located in Germany, legal action was impractical and the *LFT* forced the Phönix company

Above: Engine testing of a Brandenburg G.I in early 1917 at *Fluggeschwader* I. Dual wheels have been fitted to strengthen the landing gear.

Left: Brandenburg G.I 62.56 with landing gear collapsed. This shows the aircraft after modification in late 1917/early 1918; the pilot's cockpit was moved forward.

in Austria, which was under the same ownership as Brandenburg, to assume full financial responsibility for business dealings between Brandenburg and the Austro-Hungarian authorities.

The G.I had a bombardier/gunner in the front cockpit; the pilot and rear gunner shared the aft cockpit. The bomb load was five 50 kg bombs and four 20 kg bombs, a total of 330 kg. Once production bombers reached the front the units had to make extensive repairs and modifications well beyond that normally expected. After massive efforts only a single bombing mission by the G.I was flown before all the aircraft were stored awaiting the outcome of the dispute between Brandenburg and the *LFT*.

In late 1917 the Austro-Hungarian army requested night bombing on Italian units and the *LFT* decided to rebuild the G.I bombers in storage. The G.I was still unsatisfactory despite the upgrades applied and the G.I bombers were primarily used as twin-engine trainers in preparation for receiving Gotha G.IV(LVG) bombers from Germany.

Brandenburg G.I Production Orders

Serials	Qty	Mfr	Order Date
62.01–06	6	UFAG	July 7, 1915
62.51–56	6	Brand.	Dec. 12, 1915
62.57–77	21	Brand.	Dec. 18, 1915
62.07–12	6	UFAG	Feb. 25, 1916

Right: Brandenburg G.I 62.54 at Briest with a 70 mm Skoda cannon (or mockup?) fitted in the nose for ground attack. A newer mounting for the 70 mm Skoda was tested on G.I 62.62, but neither installation saw combat.

Above: Brandenburg 05.05 with pilot's cockpit well aft.

Left: Brandenburg G.I 62.07 has come to grief after landing gear failure. The landing gear of the G.I was one of its weaker points, but the only factory response was to install dual wheels on some production bombers.

Below: Installation of a 70 mm Skoda cannon for ground attack in Brandenburg G.I 62.54 at Briest. A co-axial machine gun above the cannon was fitted for aiming. A newer mounting for the 70 mm Skoda was tested on G.I 62.62, but neither installation saw combat. The profusion of struts made the central airframe and engine assembly strong at the expense of significant drag.

LVG G-Type Bombers

Founded in early 1910, the LVG (Luft-Verkehrs-Gesellschaft mbH) company became one of the most important aircraft production companies in Germany during the war. Virtually all LVG aircraft to reach production were two-seat reconnaissance airplanes, but LVG also built some experimental fighters and bombers that were not built in quantity for operational service.

LVG G.I

The LVG G.I was designed to the 1914 *Kampfflugzeug* (battle plane) specification that lead to the AEG K.I and G.I among other designs. Featuring a forward gunner standing upright with a clear field of fire above the wings and propellers, the LVG G.I was clearly designed to attack other aircraft in the manner of naval combat, a concept quickly shown to be faulty once airplanes engaged in actual air combat.

Made possible by the abbreviated nose, the two 150 hp Benz Bz.III engines were mounted close together. In addition, the engines rotated in opposite directions to cancel propeller torque on the aircraft. The direction of rotation was chosen to minimize asymmetric thrust in case of engine failure, assisted by the proximity of the engine to the aircraft centerline. While the design gave the forward gunner an exceptional field of fire, the profusion of struts and the gunner's upright position, fully exposed to the air stream, created excessive drag that certainly limited speed. Apparently only one example of the G.I was built in 1915. Performance was undoubtedly insufficient to intercept enemy aircraft and the design had too much drag for a successful bomber.

Above: The LVG G.I was designed as a battle plane, and a single example was built in 1915.

LVG G.III

No information is available on the LVG G.II, apparently an un-built project, and the next and last LVG bomber was the massive G.III. Unusually, the G.III was built by LVG to a Schütte-Lanz design, the Schütte-Lanz G.V, a drawing of which is in the Schütte-Lanz section. Why LVG should build a

Above & Below: The LVG G.III was designed as a night bomber; it was built in 1918.

Schütte-Lanz design is not known but undoubtedly had something to do with manufacturing capability, at which LVG excelled.

The LVG G.III was a massive, twin-engine triplane bomber with biplane tail and three vertical tail surfaces. The airframe was made of wood and the fuselage was covered with plywood for strength and a smooth surface finish. The middle wings extended outward from the engine nacelles and did not reach the fuselage. There were gun positions fore and aft; each gunner was equipped with a single flexible machine gun.

Designed as a night bomber, the G.III's triplane configuration emphasized bomb load and reliability rather than speed or altitude performance. Unfortunately, the G.III appeared to late to go into production before the Armistice and little is known about its handling qualities.

Above & Below: The triplane design of the LVG G.III emphasized bomb carrying capacity for its intended role as a night bomber. The middle wings were attached to the engine nacelles instead of the fuselage.

LVG G.III Specifications

Engines:	2 x 245 hp Maybach Mb.IV	
Wing:	Span	24.6 m
	Area	115.0 m²
General:	Length	10.25 m
	Height	3.9 m
	Empty Weight	2,960 kg
	Loaded Weight	4,100 kg
Maximum Speed:		130 kmh
Climb:	3,000 m	20 min
Duration:		5½ hrs
Armament: 2 flexible MGs		

Roland G.I

The Roland G.I was Roland's only G-type design; the 'G' designation, for *Grossflugzeug* (large aircraft), later became synonymous with twin-engine bombers. Despite the fact that all other G-types were twin-engine aircraft, the Roland G.I was a single-engine aircraft. The fuselage-mounted engine drove two pusher propellers via gears and shafts. The fact that there were only two crewmen, with the gunner handling a single flexible machine gun in the nose, indicates that the G.I was originally conceived as a *Kampfflugzeug* (battle plane), or aerial cruiser, similar in concept to the AEG K.I. The AEG K.I was fitted with two 100 hp engines; the Roland G.I had more power from its single 245 hp Maybach Mb.IV engine. Despite its array of drag-producing mounting struts for the propellers, the G.I was said to be capable of 160 km/h, a good speed for a G-type at the time. The landing gear was neatly designed and featured twin nose wheels to prevent nosing over on landing.

Roland G.I Specifications		
Engine:	245 hp Maybach Mb.IV	
Wing:	Span	30.1 m
General:	Length	15.9 m
	Empty Weight	2,750 kg
	Loaded Weight	4,300 kg
Maximum Speed:		160 km/h

The *Kampfflugzeug* concept was a failure because the lumbering aircraft could not catch faster two-seaters, and these aircraft quickly found their true role as bombers. With its single engine centrally mounted at its center of gravity, the Roland G.I was not suited to all the modifications necessary to convert it into a bomber, including adding a third crewman aft as a rear gunner. As far as is known only a single G.I was built. A G.II design was ordered from Roland at the same time the G.I design was ordered, but the G.II was not built.

Above: The Roland G.I was the only Roland G-type design to be built. Roland built only single-engine aircraft; despite having two propellers, the G.I was powered by a single 245 hp Maybach Mb.IV engine located in the fuselage as indicated by the radiators mounted on the fuselage sides. The two propellers were driven by gears and shafts. The two crewmen are in their cockpits, with the gunner demonstrating his flexible machine gun. The Roland G.I was apparently designed for the *Kampfflugzeug* (battle plane), or aerial cruiser role, which quickly proved ineffective in combat. The claimed maximum speed of 160 km/h seems optimistic given the profusion of drag-producing struts supporting the propellers.

Above & Below: Two views of the Roland G.I showing the two propellers supported by struts with the gunner demonstrating his flexible machine gun. The simple landing gear appears to be effective in reducing nose-overs.

Above: Rear view of the Roland G.I showing the complex propeller struts and drive shafts.

Rumpler Bombers

Rumpler produced a comparatively small number of bomber designs that saw modest production in keeping with their performance. In addition to at least three prototypes, 58 Rumpler production bombers of all types were built as detailed in the adjacent table.

With limited engineering resources, Rumpler eventually abandoned bomber design in favor of more critical fighters and reconnaissance airplanes.

Rumpler Bomber Production		
Type	Qty	Serials
G.I	4	15–18/15
G.II	24	106–117/15 & 119–130/15
G.III	30	300–329/16. See Note.

Note: *Idflieg* ordered 50 Rumpler G.III bombers but only 30 were built.

Rumpler 4A15

Above: The Rumpler 4A15 was in response to an *Idflieg* request for a *Kampfflugzeuge*, or battleplane. Powered by two 150 hp Benz Bz.III engines mounted as pushers, it set the basic configuration for all subsequent Rumpler bombers.

In July 1914 Rumpler was among several manufacturers asked to develop a *Kampfflugzeuge*, a battleplane or aerial cruiser. The aircraft was to have a crew of two or three with a flexible gun for the observer/gunner mounted in the front cockpit, and was to have at least 200 hp.

Rumpler's response was the 4A15, a twin-engine biplane of conventional wood, wire, and fabric construction powered by 150 hp Benz Bz.III engines mounted in pusher configuration. Each engine was mounted on the lower wing and enclosed in a nacelle housing a frontal Windhoff radiator and a 310 liter fuel tank. A gravity tank was installed above each engine. The 4A15 had a span of 18.75 m and a length of 11.8 m. The simple landing gear included a pair of wheels under each engine and another pair under the

Above & Left: The Rumpler 4A15 was the first Rumpler bomber design, and all subsequent Rumpler bombers followed its general configuration of pusher engines and simple landing gear with nose wheels to prevent nose-overs on landing. The fuel tank below the engine is clearly visible at left. The propellers were mounted on short extension shafts.

nose to prevent nose-overs on landing.

The 4A15 first flew in March 1915 and achieved a maximum speed of 135 km/h at sea level. During informal flight demonstrations the aircraft carried ten people to 3,200 m on March 15th and then carried 16 people to 1,800 m. On 22 March the aircraft reached 2,300 m in 55 minutes while carrying 12 people. However, while flying to Munich on April 10 the 4A15 was forced to make an emergency landing and was damaged. On April 17, 1915 it was destroyed following a carburetor fire. However, during its brief life the 4A15 set the basic configuration for all Rumpler bombers that followed.

Rumpler G.I

Above: This photograph shows the Rumpler G.I with the first design of fin and rudder, the most visible difference between the Rumpler G.I, military designation for 5A15, and 4A15. A gravity tank was also installed just below the upper wing; the main fuel tank for the starboard engine is visible mounted beneath the engine and forward of it. The fuel tank was located near the aircraft center of gravity so burning off fuel would not adversely impact the type's pitch stability.

Rumpler continued bomber development with the 5A15. Based on the 4A15, the 5A15 used the same basic structure and engines but had a number of refinements. The wing span was increased slightly, the shape of the vertical tail surfaces was revised, and a single gravity tank under the top wing replaced the two separate gravity tanks over the engines. The gunner in the front cockpit was given a windshield and the rear crew member now had a flexible gun.

The 5A15 first flew on September 4, 1915 and demonstrated acceptable performance, although flight testing revealed a number of improvements were needed. The test pilot, probably Friedrich Budig, then chief test pilot at Rumpler, complained that the fuel system was too complicated and the gravity tank was the sole feed to the carburetors, creating a single point of failure. The aircraft was also too tail heavy and the pusher propellers were subject to damage from debris thrown up by the landing gear. Inadequate clearance (4 cm) between the fuselage and the propeller arc sometimes caused the propellers to hit the screen protecting the rear gunner from the propeller arc.

The 5A15 passed its acceptance flight on September 16, 1915, and *Idflieg* ordered it into limited production as the Rumpler G.I, only four aircraft being built. Photographs show the production aircraft had an enlarged rudder for better directional control after an engine failure. The very small number ordered indicates that the aircraft were primarily intended for operational evaluation. One aircraft, G.16/15, was delivered to *Kampgeschwader der Obersten Heeresleitung* 1 (*Kagohl* 1) for operational evaluation, where it was flown for a time by *Leutnant* Ray, while two others, G.17/15 and G.18/15, served briefly that year with *Brieftauben Abteilung Metz*, or *B.A.M.*

Below: Rumpler G.I 15/15 in flight.

Left: This photograph shows the Rumpler G.I engine installation in more detail. The under-wing gravity tank and main fuel tank for the starboard engine are clearly visible. The frontal radiator was large and complex.

Below & Bottom: Rumpler G.I 15/15, the first of four production machines. In this front quarter view the most visible difference between the Rumpler G.I and A415 is the windshield provided for the front gunner in addition to the gravity tank installed just below the upper wing. The front gunner's flexible machine gun is also visible.

Above: Rumpler G.I 15/15 is shown moments after lifting off during take-off. This view confirms that the enlarged rudder was retrofitted to this aircraft.

Below: This front quarter view of Rumpler G.I 16/15 shows the overall workman-like design and construction. The tents in the background indicate this is an operational aircraft at its unit.

Rumpler G.I Specifications		
Engines:		2 x 150 hp Benz Bz.III
Wing:	Span	19.3 m
	Area	78.68 m²
General:	Length	11.8 m
	Height	4.0 m
	Empty Weight	1,998 kg
	Loaded Weight	2,938 kg
Maximum Speed:		145 kmh
Climb:	800m	7 min
	2000m	21 min
	4000m	120 min
Service Ceiling:		4000 m
Range:		600 km
Armament: 1 flexible MG & 150–200 kg of bombs		

Above & Below: Rumpler G.I 16/15, the second of four production machines, photographed at the Fokker factory. The Fokker V1 prototype is at far right in the photo above.

Above & Below: Rumpler G.I 16/15, the second of four production machines, photographed at the Fokker factory. The fuselage insignia is in a different location than in the photos on the facing page.

Above, Below & Bottom: Rumpler G.I 16/15, the second of four production machines. The most noticeable difference between this aircraft and the first G.I is the enlarged rudder with large aerodynamic balance for improved directional control with one engine out. The larger rudder was retrofitted to G.I 15/15. The engine cowlings were well streamlined. A black and white measurement bar is included in two photographs to enable measurements to be scaled from the photos.

Above: A Rumpler G.I on the Eastern Front with its lower starboard engine cowling removed for maintenance.

Rumpler G.I 16/15.

Rumpler G.II

In November 1915 *Idflieg* ordered 24 improved aircraft as the Rumpler G.II. The most important change was the use of more powerful, 220 hp Benz Bz.IV engines for improved speed, climb, and ceiling. The G.I airframe was used with minor modifications, but flight testing soon revealed the need for additional modifications for improved flying qualities. The prototype G.II, Rumpler internal designation 5A16, was tail heavy and insufficiently stable. Testing revealed the need for larger propellers to efficiently absorb the greater engine power, so propellers of 3.1 m diameter were fitted and the fuselage was narrowed slightly to provide sufficient clearance.

Rumpler chief test pilot Friedrich Budig determined that modifications to the lower wing near the propellers were needed, and on May 27, 1916 large cut-outs were made in the lower wing in front of the propellers on the second production

Rumpler G.II Specifications		
Engines:	2 x 220 hp Benz Bz.IV	
Wing:	Span	19.3 m
General:	Length	11.8 m
	Height	4.0 m
	Empty Weight	1,990 kg
	Loaded Weight	2,990 kg
Maximum Speed:		164 kmh
Armament: 2 flexible MGs & bombs		

aircraft. This modification greatly reduced tail heaviness as determined by Budig during a successful test flight on May 30.

Rumpler G.II 117/15, the last aircraft of the first production series, was fitted with more powerful 260 hp Mercedes D.IVa engines and larger propellers of 3.17 m diameter. *Idflieg* accepted this modified aircraft and it went on to serve with *Kagohl* 2 on the Eastern Front in the summer of 1916. No other G.II

Rumpler G.I, G.II, & G.III

179

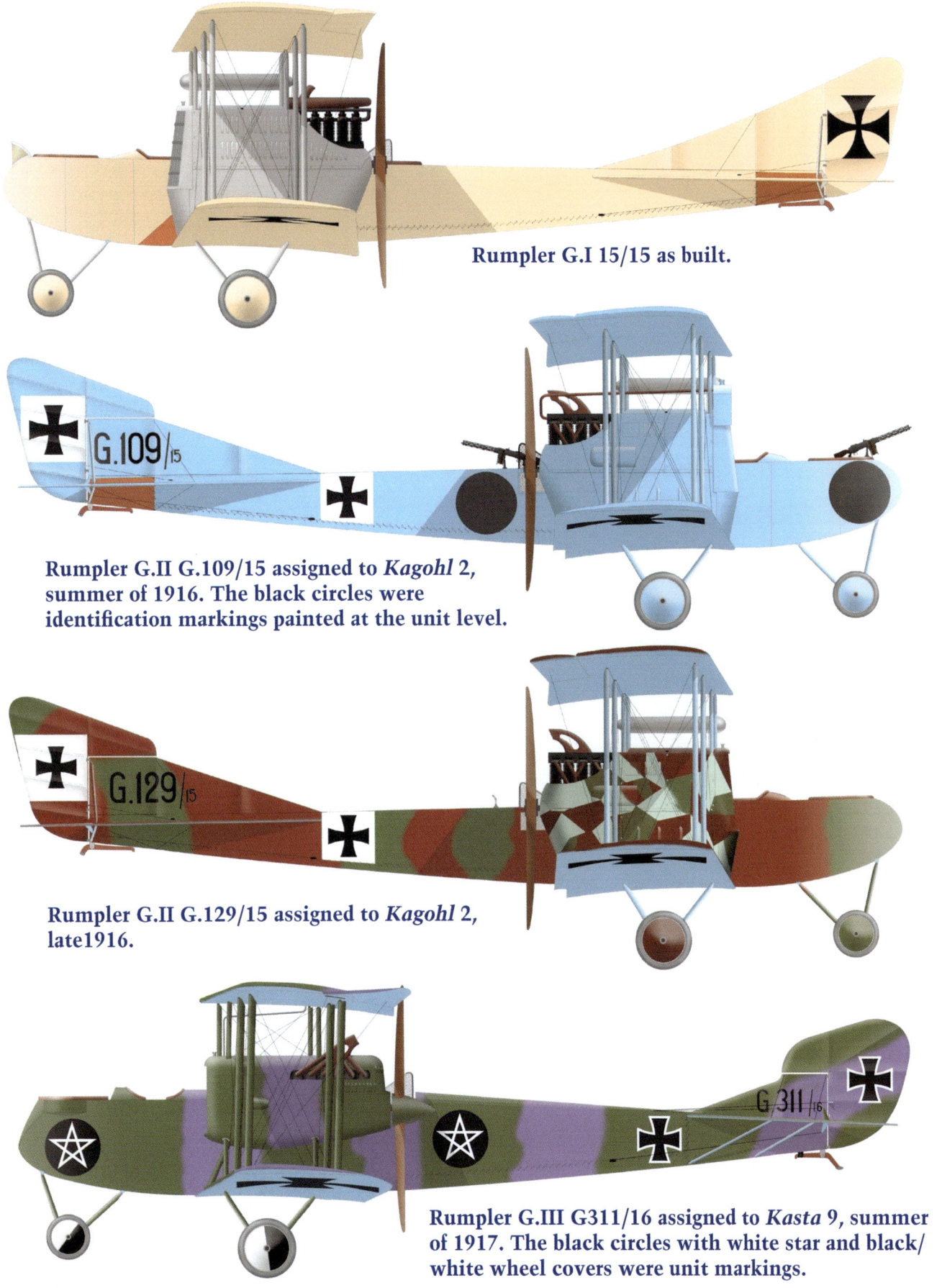

Rumpler G.I 15/15 as built.

Rumpler G.II G.109/15 assigned to *Kagohl* 2, summer of 1916. The black circles were identification markings painted at the unit level.

Rumpler G.II G.129/15 assigned to *Kagohl* 2, late 1916.

Rumpler G.III G311/16 assigned to *Kasta* 9, summer of 1917. The black circles with white star and black/white wheel covers were unit markings.

Left: The Rumpler G.II was developed from the Rumpler G.I by replacing the 150 hp Benz Bz.III engines of the G.I with more powerful 220 hp Benz Bz.IV engines. This early G.II, possibly the first G.II airframe, is virtually indistinguishable from the G.I.

Left: The more powerful Rumpler G.II was much more difficult to fly than the G.I On 27 May 1916 large cutouts were made in each lower wing beneath the propellers on Rumpler G.II 107/15, the second G.II, as shown here. On 30 May Rumpler test pilot Friedrich Budig successfully flew this aircraft, and all subsequent G.II aircraft featured this cutout.

Below: The Rumpler G.II was developed from the Rumpler G.I by upgrading the engines. This G.II, 117/15, the last aircraft of the first production batch of G.II bombers, is virtually indistinguishable from the G.I other than the barely visible cutout of the lower wing trailing edge. This is also the one G.II modified to have 260 h.p. Mercedes D.IVa engines, but the photo is not clear enough to show which engine type is fitted.

aircraft used this engine, but it was a harbinger of things to come in the Rumpler G.III.

The Rumpler G.II saw much of its service with *Kagohl* 2. Originally formed from the *Brieftauben Abteilung Metz*, or *B.A.M.*, in December 1915, *Kagohl* 2 had six *Kampfstaffeln*, *Kastas* 7–12. Five of these had single-engine C-type aircraft, but one was equipped with the Rumpler G.II and AEG G.III. *Kagohl* 2 was transferred to the Eastern Front in July 1916 and was eventually stationed at Lasnaja aerodrome. *Kagohl* 1, originally formed from the *Brieftauben Abteilung Ostend*, or *B.A.O.*, was also transferred to the Eastern Front in June 1916, and had some Rumpler G.II aircraft along with other G-types. In September 1916 *Kagohl* 1 was transferred to northern Bulgaria where it flew a number of bombing missions against Romanian troops. *Kagohl* 2 was transferred back to the Western Front in October 1916, followed by *Kagohl* 1 in May 1917.

The Rumpler G.II bombers were better armed than the C-types and on the Eastern Front were used as long-range reconnaissance aircraft and as escorts for the smaller C-types on bombing raids.

Above: The most visible differences between Rumpler G.II, 117/15 and the earlier Rumpler G.I are the cutouts in the lower wing trailing edge and the white backgrounds for the iron cross insignia. The screens to protect the rear gunner from the propellers are also visible.

Right: Rumpler G.II G.107/15 is shown in flight over the Eastern Front while serving with *KG* 2. The light overall finish still contrasts somewhat with the white backgrounds for the iron cross insignia. Both guns are visible as is the slight sweep-back of the wings.

Above: In contrast to earlier Rumpler bombers with overall light finish, Rumpler G.II G.129/15, the next to last production G.II, carries an interesting camouflage scheme, with the camouflage on the engine nacelles painted a different pattern than the fuselage. Despite the distinctive camouflage, only standard factory markings and insignia are visible. Because the factory serial number is clearly visible, the camouflage on wings and fuselage was probably painted at the factory. The straps for securing the bombs under the fuselage are clearly visible, as is the lower wing trailing edge cutout. The summer scene was probably taken at KG2 on the Eastern Front.

Below: This close-up view of Rumpler G.II G.129/15 shows its interesting camouflage scheme and engine nacelles in more detail. The camouflage appears to be sprayed on the fuselage and brush-painted on the nacelles, indicating the nacelle camouflage may have been applied at the unit. The screens to protect the gunner from the propellers are clearly shown. Although not clear, there appears to be segmented camouflage on the bottom of the upper wing.

Above: Rumpler G.II G.109/15 of *KG*2 damaged at Lasnaja, one of two fields *KG*2 is known to have used on the Eastern Front, where *KG*2 served from mid-July to October 1916. The black circle marking is distinctive. The radiators are very different than those fitted to most other Rumpler G.II bombers, but the reason for this modification is unknown.

Below: Then *Unteroffizier* Gustav Seitz poses in front of Rumpler G.II G.110/15 of *Kampfstaffel* 9 of *KG*2 at Kowel after receiving the Iron Cross 2nd Class. Again the radiators are distinctly different than those fitted to most other Rumpler G.II bombers.

Rumpler G.III

The Rumpler G.II had proven itself moderately successful and *Idflieg* placed an order for the proposed Rumpler G.III in September 1916. Retaining the basic size and configuration of the earlier G.II, the G.III was a new design. Although wingspan was the same as the earlier types, wing area was reduced by a smaller lower wing. The engine nacelles were more streamlined and more powerful 260 hp Mercedes D.IVa engines were fitted in place of the 220 hp Benz engines of the G.II. To reduce the danger of fire the fuel tanks were now mounted in the fuselage instead of in the nacelles. Unfortunately, the weight of the new aircraft increased substantially due to increases in the empty weight and also the designed payload.

Flight testing of the new prototype, factory designation 6G2, began in December 1916 and quickly revealed unsafe flying qualities. Like the earlier designs, the G.III prototype was tail heavy in flight. To solve this problem the horizontal stabilizer was raised and the fuselage near the plane of the propellers was modified to improve airflow, and these changes greatly alleviated the tail-heaviness. Chief test pilot Friedrich Budig stated the revised G.III was more stable and had better flying qualities than the G.II. However, adding the rear gunner's gun and ammunition again caused tail-heaviness, and the upper wings had to be moved rearward to restore pitch stability and trim.

Rumpler G.III Specifications		
Engines:	2 x 260 hp Mercedes D.IVa	
Wing:	Span	19.3 m
	Area	78.68 m²
General:	Length	12.0 m
	Height	4.5 m
	Empty Weight	2,365 kg
	Loaded Weight	3,620 kg
Maximum Speed:		150 kmh
Climb:	3000m	22 min
Service Ceiling:		5000 m
Range:		660 km
Armament: 2 flexible MGs & 225 kg of bombs		

The G.III that arrived at the front in December 1916 was the aircraft assigned to *Kagohl* 2, which had used the earlier G.II. Some G.III bombers served into at least March 1918 with *Kampfstaffel* 9 of *Kagohl* 2.

The final Rumpler bomber prototype, factory designation 6G4, had increased wing span and tractor-mounted four-bladed propellers. The first test flight on 18 February 1918 indicated marginal flight characteristics when the aircraft swung to the right on takeoff and the pilot, Budig, was barely able to correct it. With extensive modifications necessary to improve its poor flying qualities and higher priority programs underway, further bomber development was abandoned.

Below: This Rumpler G.III shows its camouflage to advantage. The upper surfaces appear to be sprayed in two colors while the rudder is painted a single dark color and undersurfaces, struts, nacelles, and wheel covers appear to be in a single light shade. Propeller spinners are fitted, showing attention to streamlining details.

Above: This photograph of Rumpler G.III G.311/16 of *Bogohl* 2 is especially interesting because, unlike most G.III photos, it shows unit markings on the nose and the wheel covers are painted half dark and half light; black and white? The aircraft is loaded for a bombing mission with P.u.W. bombs under the fuselage along with a square *Kastenbombe* for blast effect.

Above: This front view of a Rumpler G.III emphasizes its clean lines with great attention to streamlining, a Rumpler hallmark.

Right: The front gunner's cockpit of the Rumpler G.III appears much more compact than the rear gunner's position despite housing a lot of equipment.

Above: Rumpler G.III photographed while serving with an operational unit.

Left: This front quarter view of a Rumpler G.III apparently at an operational unit shows its camouflage and clean lines to advantage. Interestingly, the rudder appears to be in two colors like the rest of the upper surfaces. The photo gives the impression of being of the same aircraft as the photo below, except the rudder appears to be in a single dark color. Are these two different aircraft or is this a trick of the light?

Right: This rear quarter view of a Rumpler G.III apparently at an operational unit shows that the propeller spinners were not fitted. The rear gunner's gun is fitted and his cockpit appears very roomy.

Schütte-Lanz G.I

Schütte-Lanz is mainly known for building airships, but also built a series of airplane prototypes, none of which reached production. One bomber, the Schütte-Lanz G.I, was built and several others were designed but not built, although the Schütte-Lanz G.V was built by LVG as the LVG G.III.

The Schütte-Lanz G.I was designed and built to the original *Kampfflugzeug* specification of 1914; it was originally assigned to the "K" category but became the G.I when the G-category replaced it. The fuselage had a five-sided cross section covered with plywood. Two 160 hp Mercedes D.III engines were fitted in a pusher configuration, common to most German two-engine aircraft of the time. As shown in the bottom photo, the propellers were fitted to long extension shafts to clear the wing.

The gunner had a single flexible machine gun in the front cockpit. The configuration did not look balanced; a small rudder without fixed fin was fitted but the horizontal stabilizers were nearly half the length of the fuselage. The G.I was too slow to intercept hostile aircraft and the propeller extension shafts and their supports were heavy and created excessive drag; only one aircraft was completed.

Schütte-Lanz G.I Specifications

Engines:		2 x 160 hp Mercedes D.III
Wing:	Span	22.0 m
	Area	100.0 m²
General:	Length	12.0 m
	Height	4.2 m
	Empty Weight	1,850 kg
	Loaded Weight	3,100 kg
Maximum Speed:		125 km/h
Duration:		6 hours
Armament: 1 flexible MG		

Above: Designed as a battle plane, only one Schütte-Lanz G.I was built.

Right: The heavy, drag-creating propeller extension shafts and their supports, clearly seen in this view, were the result of determining to use a pusher configuration and did nothing to enhance performance or load-carrying capability.

Above: This view shows the unusual cross-section of the Schütte-Lanz G.I fuselage.

Drawing for the proposed Schütte-Lanz G.II/G.III that was not built. The main difference from the G.I appears to be the streamlined engine cowlings.

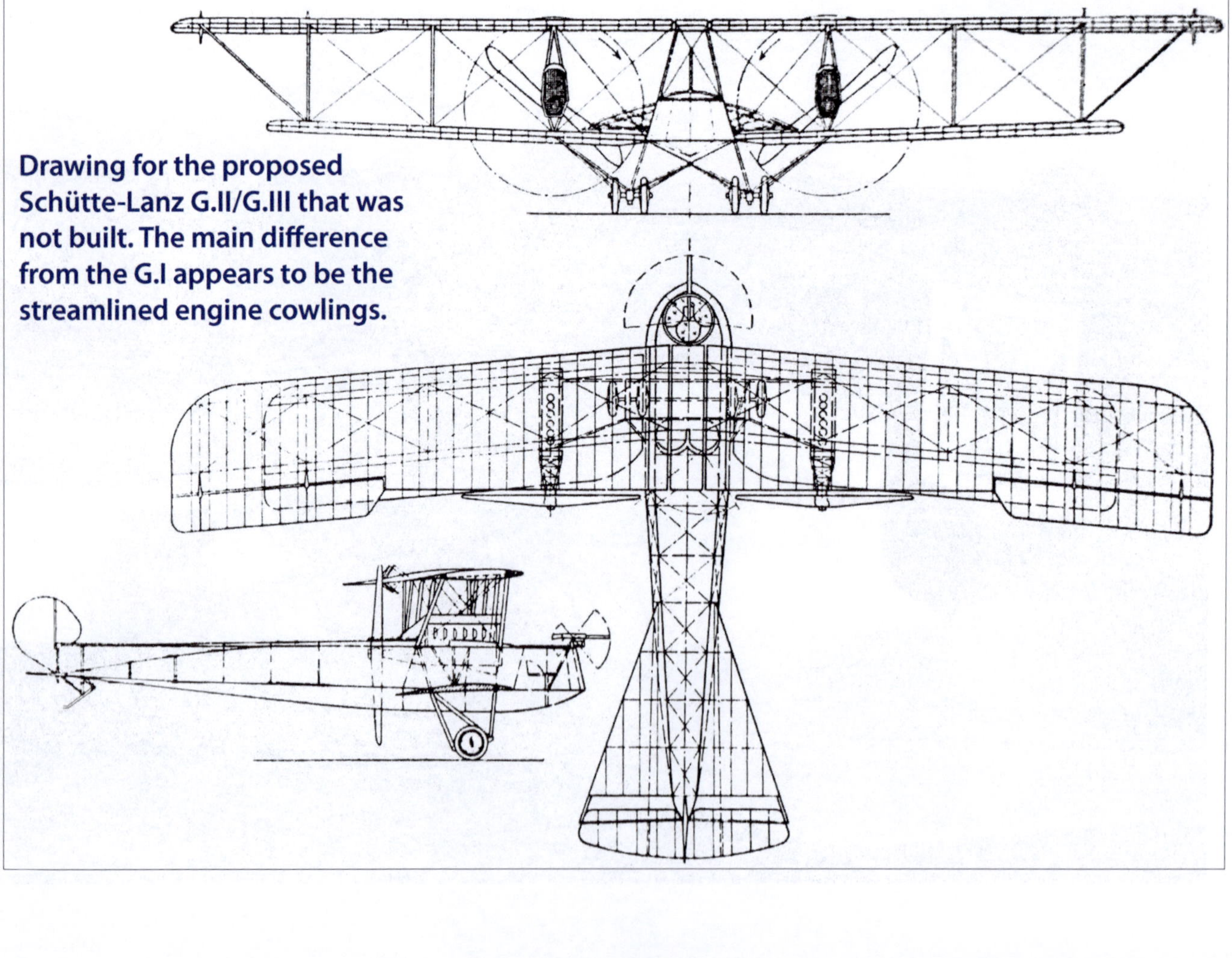

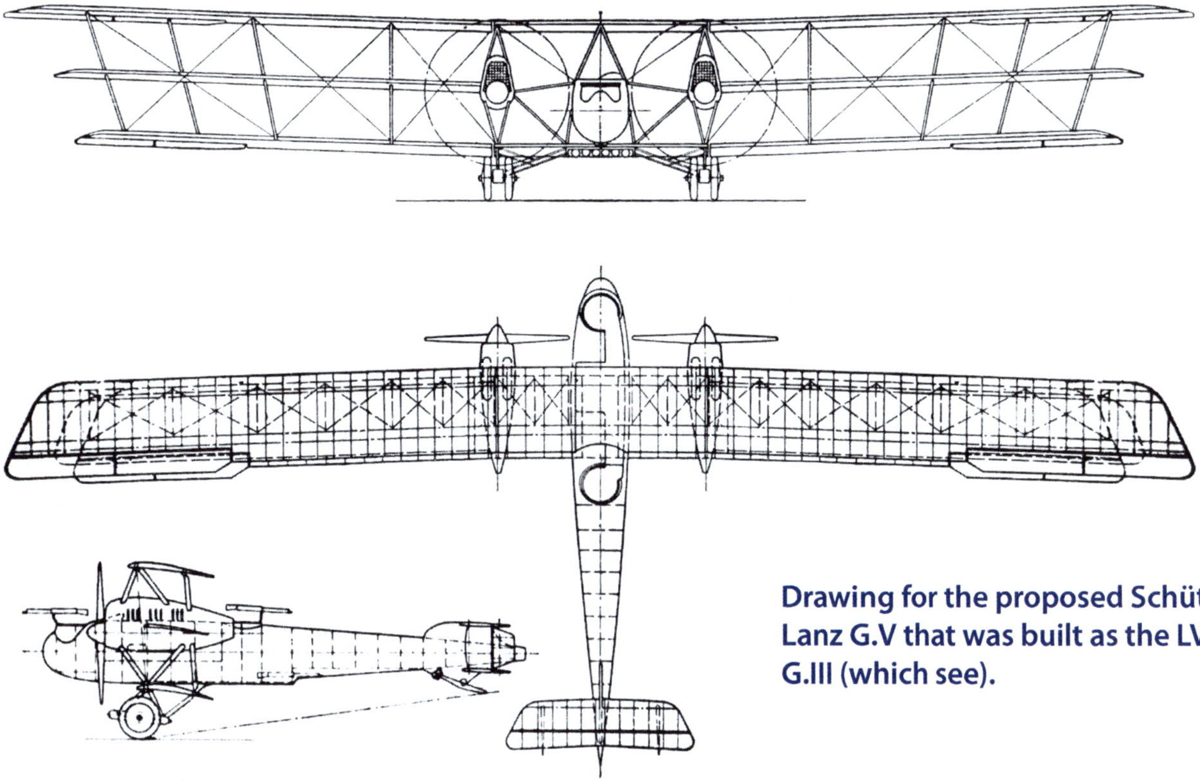

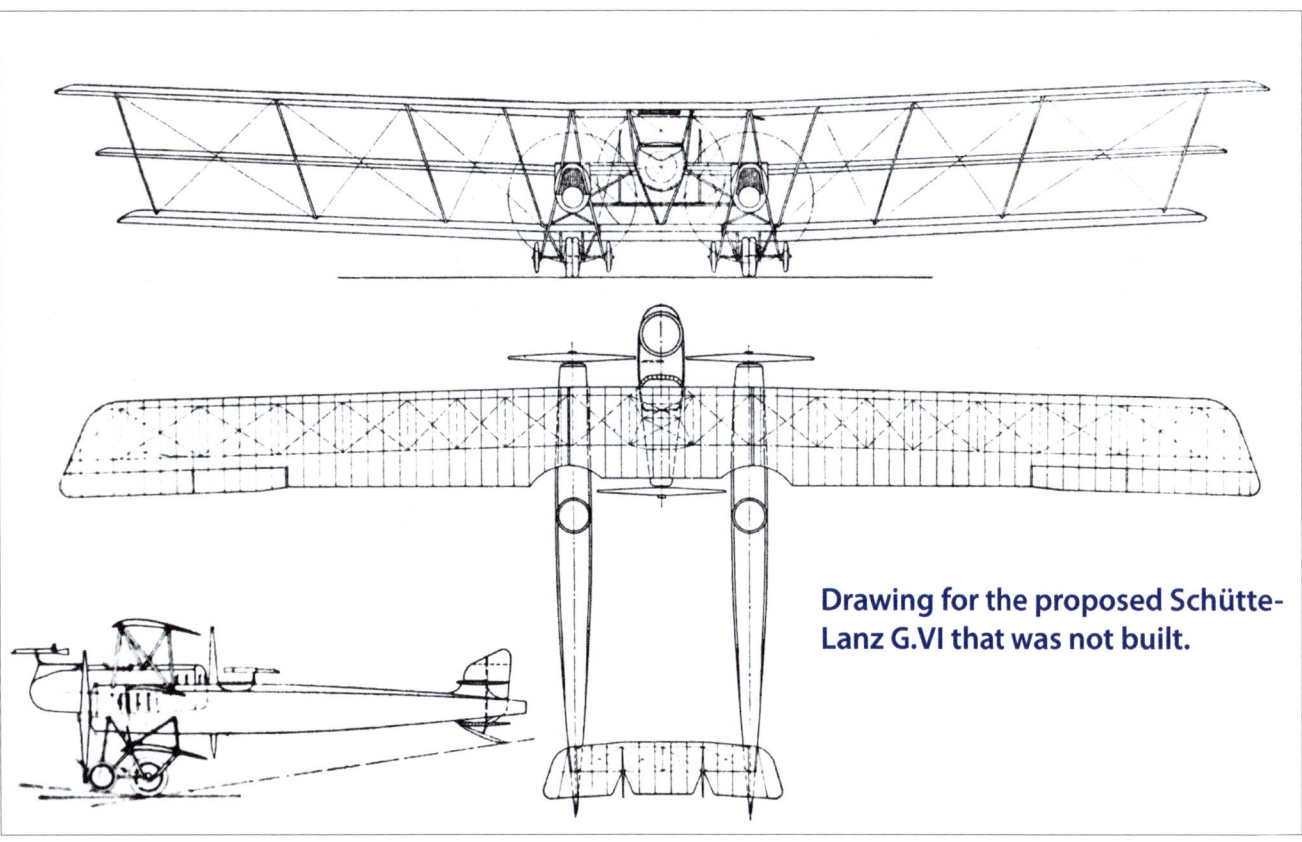

Drawing for the proposed Schütte-Lanz G.V that was built as the LVG G.III (which see).

Drawing for the proposed Schütte-Lanz G.VI that was not built.

SSW L.I

The SSW L.I was the only bomber built in *Idflieg's* new L-category, a type intermediate between the G and R types. Initially designated the G.III (the G.I and G.II remaining un-built projects), six G.III bombers, (894–899/17) were ordered in October 1917. The type was re-designated L.I in April 1918. The first bomber, L.897/17, was completed in June 1918 and made its initial flight on August 5. It was destroyed in a landing accident a few weeks later. Two more L.I bombers were completed, 898/17 in October 1918 and 899/17 in February 1919.

Seimens-Schuckert L.I Specifications

Engines:	3 x 240 hp Maybach Mb.IV	
Wing:	Span, Upper	32.0 m
	Wing Area	169 m²
General:	Length	14.65 m
	Empty Weight	4,400 kg
	Loaded Weight	6,400 kg
Maximum Speed:		125 km/h
Endurance:		5.5 hrs.

Above: The SSW L.I was the only aircraft completed to the new L-type category.
Below: The SSW L.I was a large, powerful aircraft that used the tri-motor Caproni-type configuration.

Above: The SSW L.I had a gunner behind the wing in each fuselage boom, and they could fire downward through a Gotha-type tunnel in the tail boom to defend against attacks from directly below.

Above: The SSW L.I 898/17, which was probably the aircraft completed in October 1918. Powered by three 240 hp Maybach Mb.IV engines, the L.I was armed with three flexible machine guns.

The L.I followed the configuration of the Italian Caproni bombers with two fuselage booms and three engines, one in the rear of the short fuselage and two others in the front of the booms. Power was provided by three 240 hp Maybach Mb.IV engines. Each boom carried a gun position and the gunners could fire downward through Gotha-type fuselage tunnels. A third gun position was in the fuselage nose. The Armistice ended development of the L.I before it could be tested in combat

German G-Types in Retrospect

Originally developed for the flawed battle plane concept, German twin-engine aircraft soon found their true role as bombers as a result of combat experience. A wide variety of designs were built by many manufacturers, and the most successful were those built by AEG, Gotha, and Friedrichshafen. Of these the Gotha bombers had the worst flight characteristics due to being tail heavy, but were also by far the most famous due to bombing London in daylight, an attack that shocked the world and made the name 'Gotha' notorious.

Nearly all G-types were conventional biplanes with engines mounted between the wings, typically in pusher configuration, with gunners fore and aft. Exotic structures and aerodynamics were avoided. The LVG G.III twin-engine triplane and SSW L.I tri-motor biplane prototypes arrived too late to be produced in quantity.

Eventually the powerful and reliable 260 hp Mercedes D.IVa six-cylinder engine predominated. More powerful engines were generally not available and special high-altitude engines were tested but not used because, by the time those were available, the G-types were used solely for night bombing where high altitudes were not employed. For night bombing the emphasis was on reliability, safe flight characteristics, and heavy bomb load.

Apart from the spectacular Gotha raids on the UK, G-types spent their careers bombing tactical targets close behind Allied lines. As defenses improved these raids were made at night and the aircraft that performed them were largely anonymous. The raids were a continual problem for the Allies, who required the surrender of all night bombers as part of the terms of the Armistice. Considering the damage the night bombers caused, the Allies were amazed at the small number of bombers turned over and at first refused to believe this small number of aircraft could have been so effective, which was a great testament to these aircraft and their crews.

Bibliography

Books

Borzutzki, Siegfried, *Flugzeugbau Friedrichshafen GmbH*, Verlag Markus Burbach, 1993.

Gray, Peter, and Thetford, Owen, *German Aircraft of the First World War*, second revised edition, New York: Doubleday & Company, Inc., 1970.

Grosz, Peter M., *AEG G.IV*, Albatros Publications, Berkhamstead, 1995.

Grosz, Peter M., *Fdh G.III–IIIa*, Albatros Publications, Berkhamstead, 1997.

Grosz, Peter M., *Gotha!*, Albatros Publications, Berkhamstead, 1994.

Grosz, Peter M., *Gotha G.I*, Albatros Publications, Berkhamstead, 2000.

Grosz, P.M., Haddow, G., & Schiemer, P., *Austro-Hungarian Army Aircraft of World War One*, Flying Machines Press, 1993.

Herris, Jack, *Gotha Aircraft of WWI*, Aeronaut Books, 2013.

Herris, Jack, *Rumpler Aircraft of WWI*, Aeronaut Books, 2014.

Herris, Jack, *Siemens-Schuckert Aircraft of WWI*, Aeronaut Books, 2014.

Owers, Colin A., *Late Gotha Bombers*, Albatros Productions, Ltd., 2010.

Articles

Chapman, Peter, "The Albatros G.II & G.III", *Over the Front* Vol.29 No.1, Spring 2014, p.12–26.

Chapman, Peter, & Ansell, Richard, "The Rumpler G.I–III", *Over the Front* Vol.26 No.4, Winter 2011, p.328–349.

Grosz, Peter M., "Daimler Aircraft of World War One" *Over the Front* Vol. 21, No.3, Fall 2006, p.250–280.

Grosz, Peter M., "Frontbestand" *WW1 Aero* No.107, Dec. 1985, p.60–66.

Grosz, Peter M., "Frontbestand" *WW1 Aero* No.108, Feb. 1986, p.66–69.

Index

Name	Page
Aschoff, Walter	128
August, Duke Ernst	13
Beckmann, von, *Rittmeister*	84
Berthold, Rudolph	12, 13
Brandenburg, Ernst	122
Budig, Friedrich	172, 178, 180, 184
Burkhardt, Hans	103, 106, 108, 121, 142, 157
Bülow, Walter von	17
Chainat, *Adj.*	112
Christl, *Vzfw.*	83
Diemer, Fritz	35
Dorr, *Uffz.*	84
Doerstlingen, *Oblt.*	93
Flathow, *Lt.*	131
Friedel, *Major*	99
Gaiser	60
Grailsheim, von, *Oblt.*	84

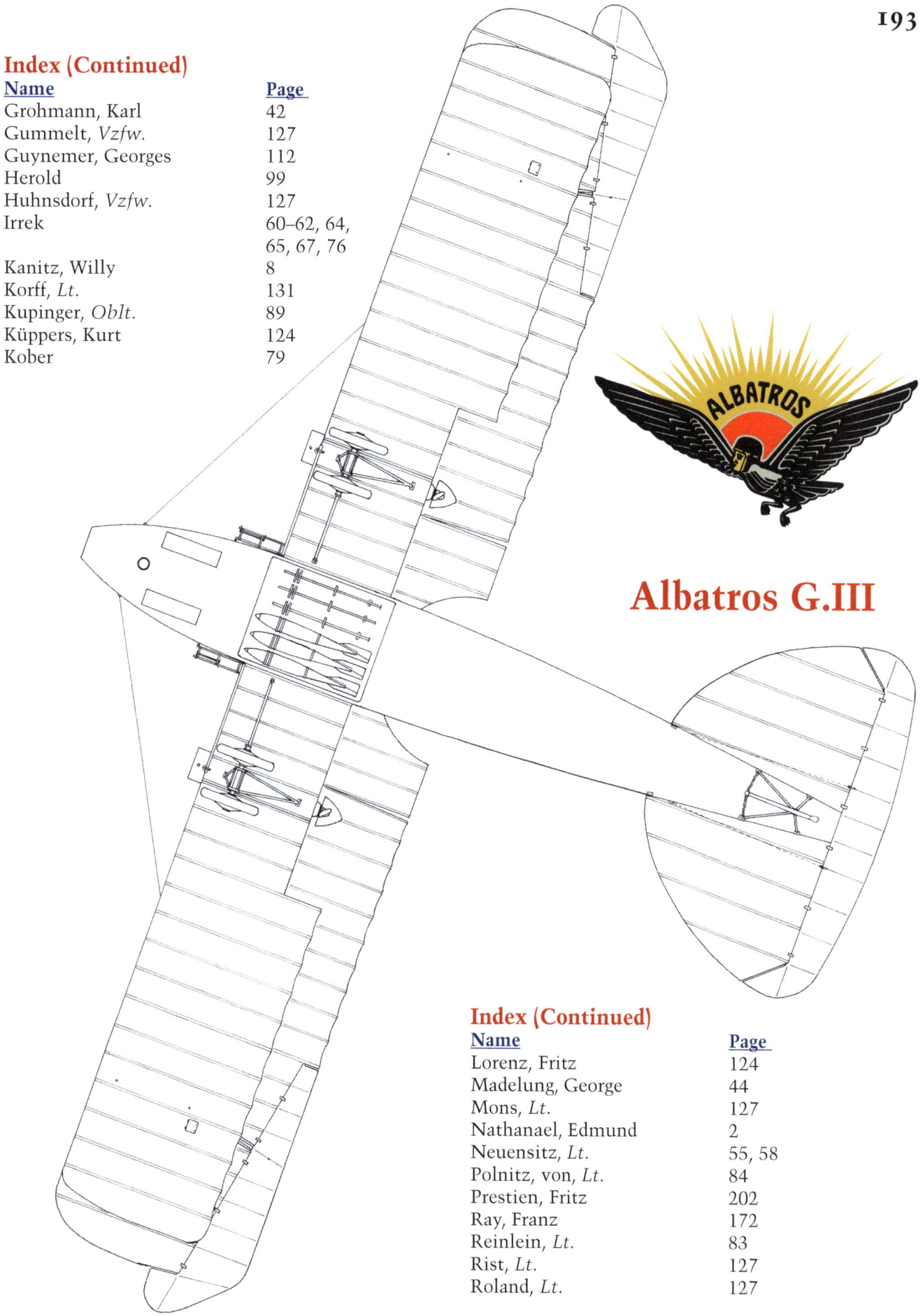

Index (Continued)

Name	Page
Grohmann, Karl	42
Gummelt, *Vzfw.*	127
Guynemer, Georges	112
Herold	99
Huhnsdorf, *Vzfw.*	127
Irrek	60–62, 64, 65, 67, 76
Kanitz, Willy	8
Korff, *Lt.*	131
Kupinger, *Oblt.*	89
Küppers, Kurt	124
Kober	79

Albatros G.III

Index (Continued)

Name	Page
Lorenz, Fritz	124
Madelung, George	44
Mons, *Lt.*	127
Nathanael, Edmund	2
Neuensitz, *Lt.*	55, 58
Polnitz, von, *Lt.*	84
Prestien, Fritz	202
Ray, Franz	172
Reinlein, *Lt.*	83
Rist, *Lt.*	127
Roland, *Lt.*	127

Albatros G.III

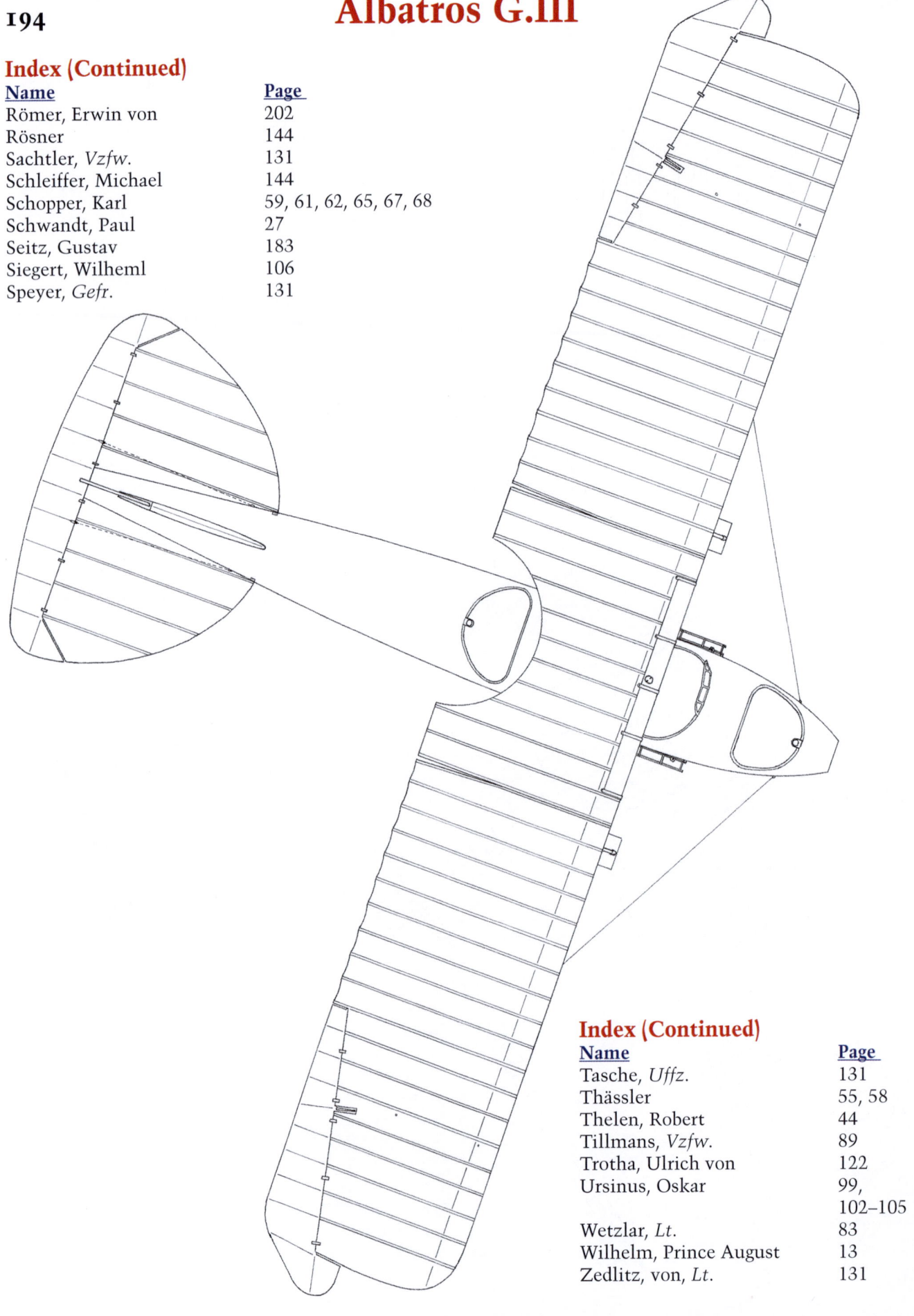

Index (Continued)

Name	Page
Römer, Erwin von	202
Rösner	144
Sachtler, *Vzfw.*	131
Schleiffer, Michael	144
Schopper, Karl	59, 61, 62, 65, 67, 68
Schwandt, Paul	27
Seitz, Gustav	183
Siegert, Wilheml	106
Speyer, *Gefr.*	131

Index (Continued)

Name	Page
Tasche, *Uffz.*	131
Thässler	55, 58
Thelen, Robert	44
Tillmans, *Vzfw.*	89
Trotha, Ulrich von	122
Ursinus, Oskar	99, 102–105
Wetzlar, *Lt.*	83
Wilhelm, Prince August	13
Zedlitz, von, *Lt.*	131

Albatros G.III

Rumpler G.I & G.II

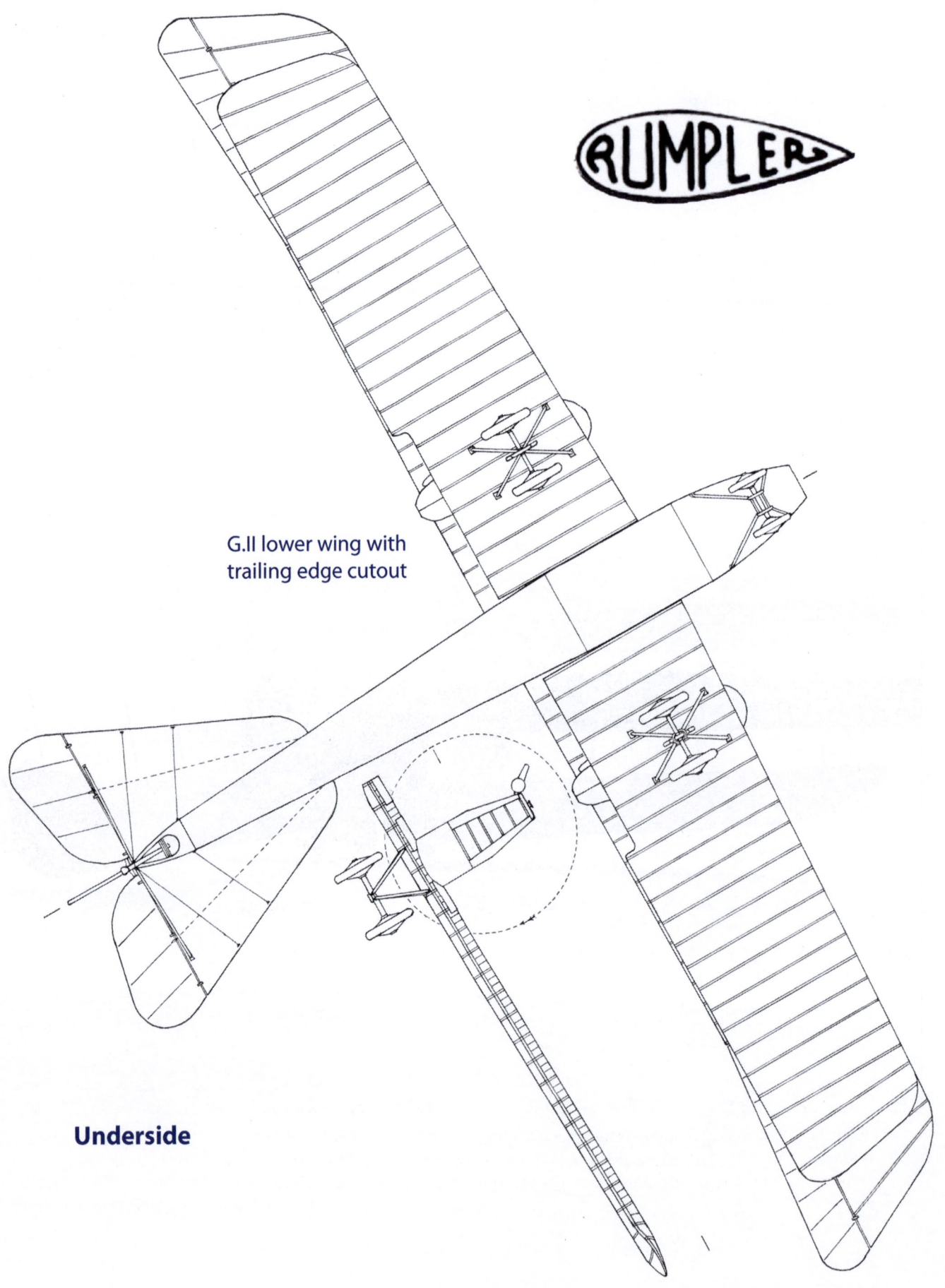

G.II lower wing with trailing edge cutout

Underside

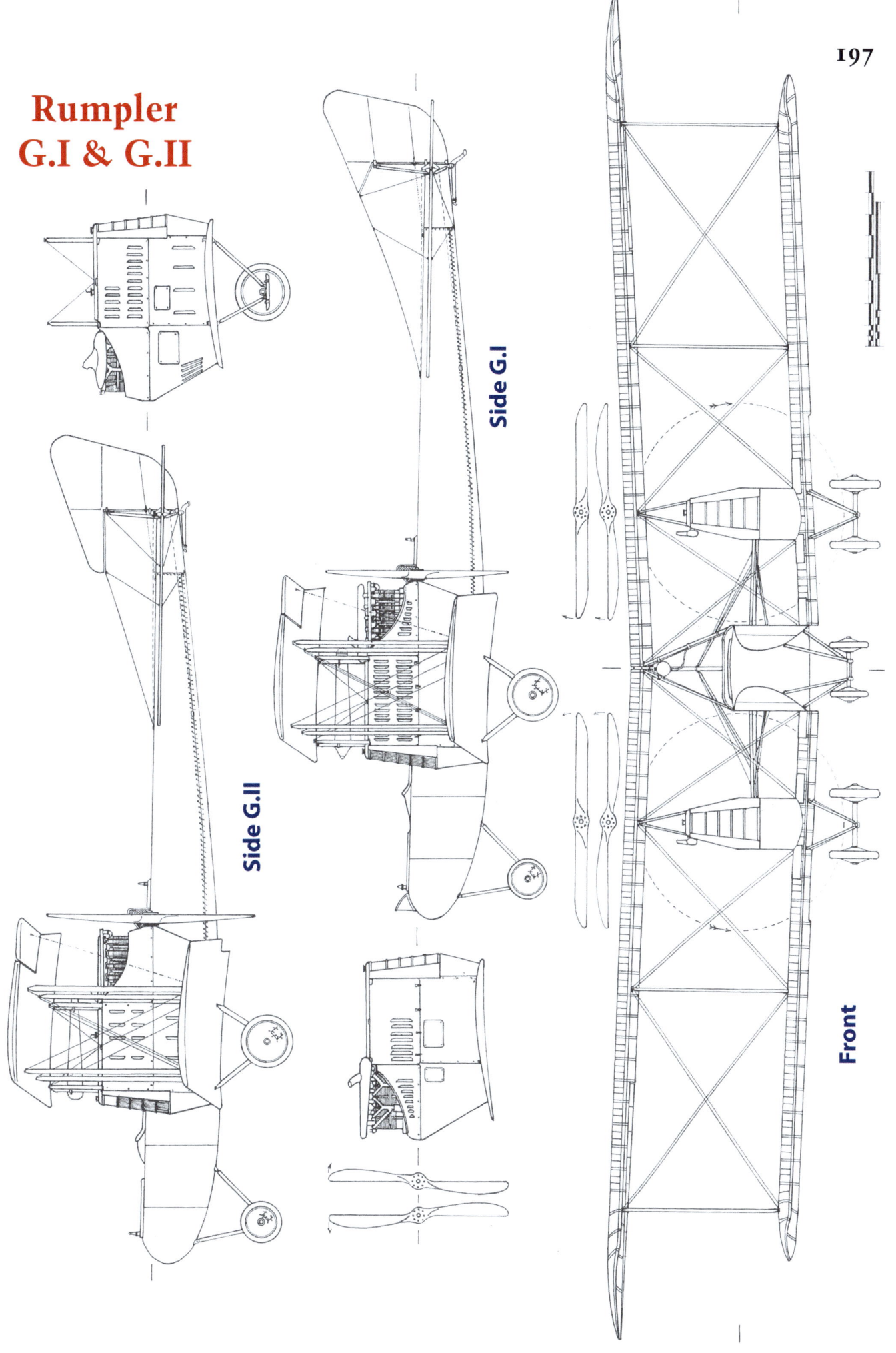

Rumpler G.I & G.II

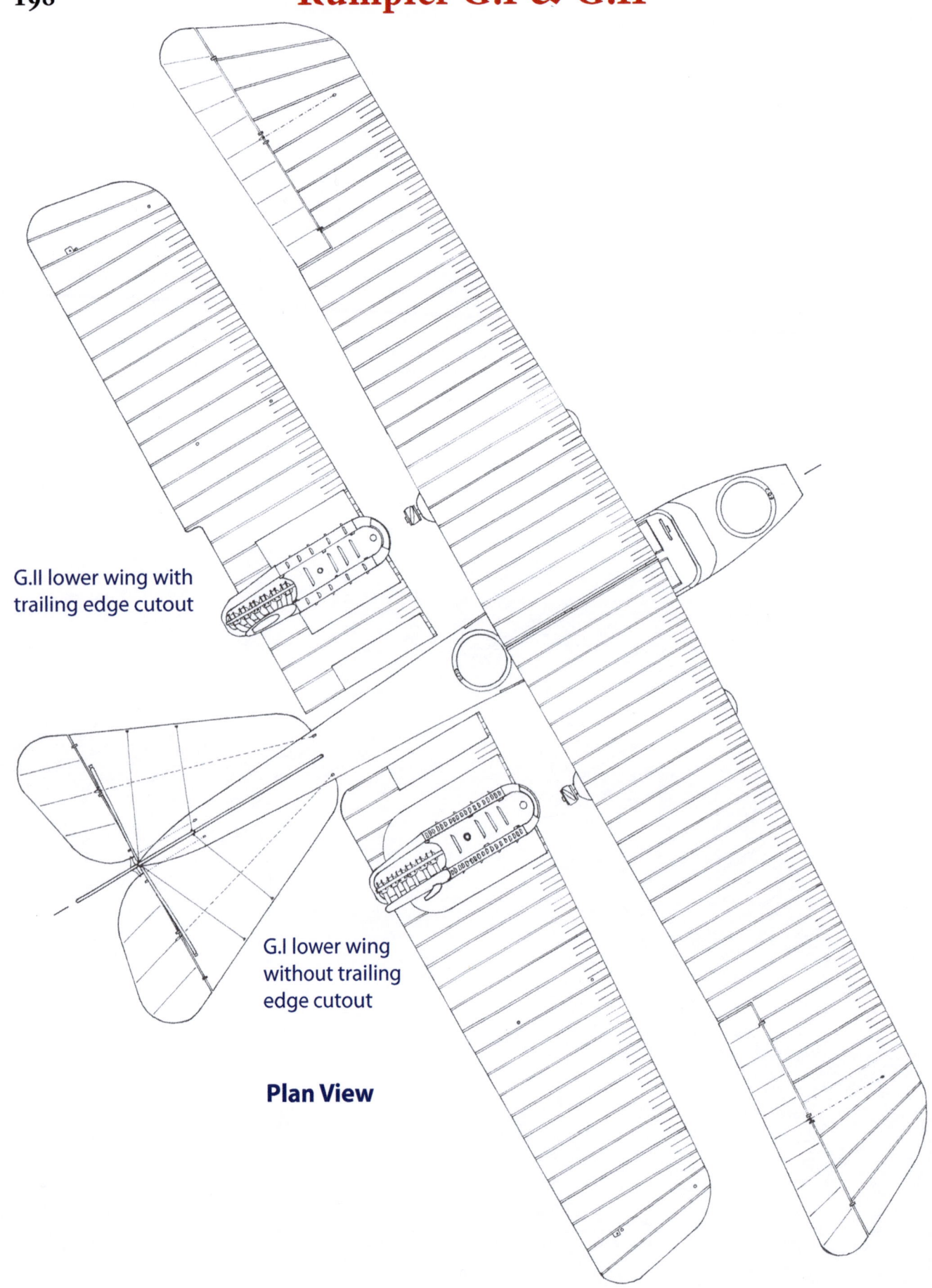

G.II lower wing with trailing edge cutout

G.I lower wing without trailing edge cutout

Plan View

Rumpler G.III

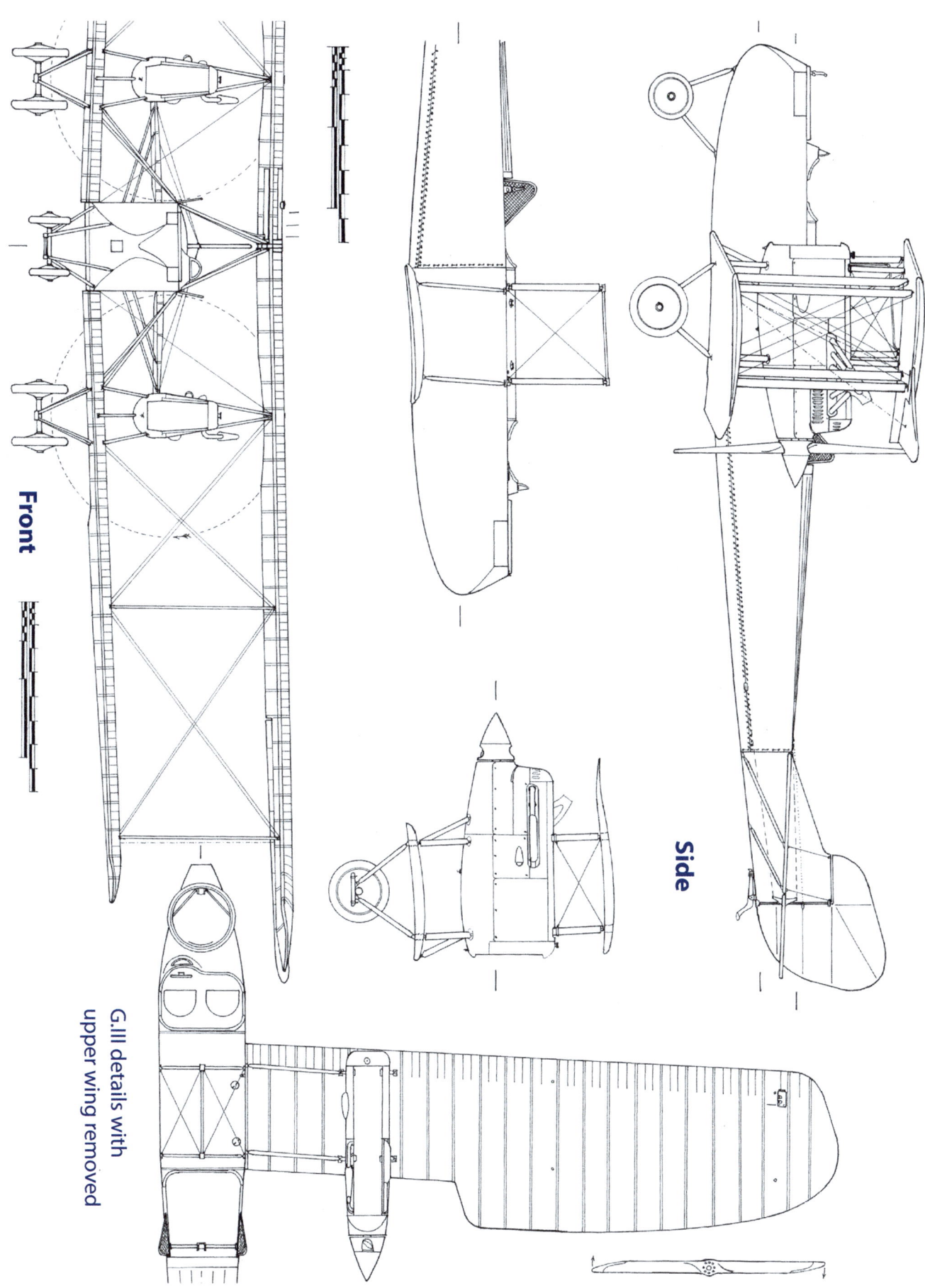

Rumpler G.III

Plan View

Rumpler G.III

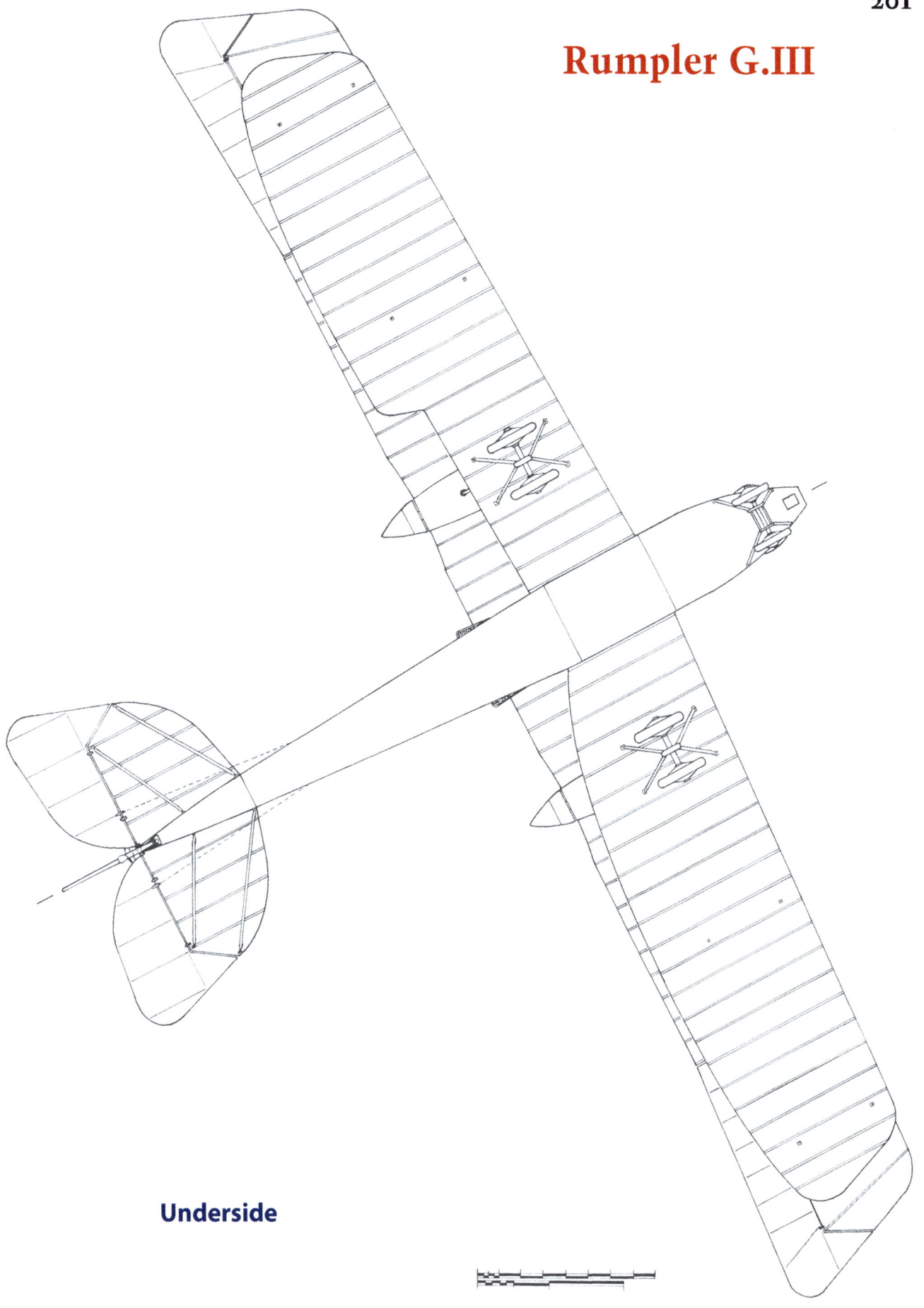

Underside

Afterword

Fighter pilots have received far more attention than bomber pilots for various reasons, but there have always been bomber heroes. Here we are pleased to honor two of them.

Left: Taken at Hudowa in April 1917, this photo shows *Hptm.* Erwin von Römer (tallest, in the center) in front of a Gotha G.II. Von Römer commanded a *Kampfstaffel* of *KG*1 where he earned the Saxon Albert Order Knight 1st Class with Swords. Von Römer also served with *Kagohl* 4. His other awards included the Reuss Honor Cross 3rd Class with Swords. Kluge, the gunner, is 3rd from left and Kempf, the pilot, is 5th from left. The other men are mechanics and ground crewmen.

Right: *Hptm.* Fritz Prestien stands in the cockpit of this Rumpler G.II serving with *Kagohl* 2; the gunner is von Hacheburg. Prestien was awarded the Knight 2nd Class with Swords of the Ducal Saxe-Ernestine House Order on Aug. 28, 1915 for his work as an *Oblt.* in *FFA* 66. At far right is *Lt.* Franz Ray, who went on to score 17 victories as a fighter pilot. On Feb. 4, 1916 Prestien was put in charge of *Kampfstaffel* 7 of *Kagohl* 2, but was severely wounded in a crash with that unit on March 12 and as a result did not return to combat until May 1918. He received numerous other awards including the Bavarian Military Merit Order 4th Class with Swords, Iron Cross 1st Class, Turkish War Medal, etc.

Printed in Great Britain
by Amazon